PEOPLES AND CULTURES OF THE MIDDLE EAST

DANIEL G. BATES
Hunter College, CUNY

AMAL RASSAM
Queens College, CUNY

PRENTICE-HALL, INC., ENGLEWOOD CLIFFS, NEW JERSEY 07632

Library of Congress Cataloging in Publication Data

BATES, DANIEL G.

Peoples and cultures of the Middle East.

Includes bibliographical references and index.

1. Near East—Social life and customs. I. Rassam, Amal. II. Title.

DS57.B34 1983 956 82-21547

ISBN 0-13-656793-2

TO SOPHIA, NINA, AND MURAT WITH LOVE

Editorial supervision and interior design: Serena Hoffman
Cover design: Diane Saxe
Cover art: Freer Gallery of Art, Washington, D.C.
Manufacturing buyer: Ron Chapman

PRINTED IN THE UNITED STATES OF AMERICA

10 9 8 7 6

ISBN 0-13-656793-2

PRENTICE-HALL INTERNATIONAL, INC., *London*
PRENTICE-HALL OF AUSTRALIA PTY. LIMITED, *Sydney*
EDITORA PRENTICE-HALL DO BRASIL, LTDA., *Rio de Janeiro*
PRENTICE-HALL CANADA INC., *Toronto*
PRENTICE-HALL OF INDIA PRIVATE LIMITED, *New Delhi*
PRENTICE-HALL OF JAPAN, INC., *Tokyo*
PRENTICE-HALL OF SOUTHEAST ASIA PTE. LTD., *Singapore*
WHITEHALL BOOKS LIMITED, *Wellington, New Zealand*

CONTENTS

Preface vii

Introduction ix

Chapter One
The Setting: Human Geography and Historical Background 1

THE GEOLOGICAL FOUNDATIONS 3
CLIMATIC REGIMES AND WATER 6
POPULATION AND SETTLEMENT PATTERNS 12
PREHISTORIC PATTERNS OF ADAPTATION 14
THE RISE OF CIVILIZATION 20
THE CULTURAL HERITAGE OF EARLY HISTORY 23
THE ISLAMIC CONQUEST AND AFTER 25

Chapter Two
Islam: The Prophet and the Religion 29

THE SCOPE OF ISLAM 30
THE RISE OF ISLAM 32
MUHAMMAD, THE MAN AND HIS PROPHECY 34
UMMA: THE ORIGINS OF ISLAMIC POLITY 39
THE REALM OF ISLAM AFTER MUHAMMAD 41
ISLAM AS FAITH 43
DUTIES AND RITUAL IN ISLAM 44
SOURCES OF ISLAMIC LAW 52
RELIGIOUS LEADERSHIP: THE *'ULAMA* 54

Chapter Three
Islam as Culture, Islam as Politics 59

SCHISMS IN ISLAM: THE SHI'A 60
SHI'A BELIEFS: A CULTURE OF DISSENT 62
THE SHI'A CLERGY 65
SECTS WITHIN SHI'ISM 67
ISLAMIC MYSTICISM: THE SUFI WAY 69
SUFI ORDERS AND BROTHERHOODS 72
POPULAR BELIEFS: *BARAKA* AND "SAINTS" 73
SUFISM AND MASS MOVEMENTS 75
THE WAHABIS OF ARABIA 76
THE SHADHILIYA ORDER OF EGYPT 77
THE SHABAK OF IRAQ 78
ISLAM AS CULTURE 81

Chapter Four
Communal Identities and Ethnic Groups 83

ETHNICITY: A THEORETICAL FRAMEWORK 84
RACE 88
LANGUAGE 89
RELIGION 92
NON-MOSLEM CONFESSIONAL COMMUNITIES 93
LEBANON AND THE MARONITES 99
THE ARMENIANS 102
MOSLEM-DERIVED SECTARIAN COMMUNITIES 104
REGIONAL ETHNIC GROUPINGS 106

Chapter Five
Pastoralism And Nomadic Society 107

THE ECOLOGICAL BASIS FOR PASTORALISM 108
ANIMALS AND NOMADISM 110
PASTORALISM AND MARKET RELATIONS 113
POLITICS AND NOMADIC PASTORALISM 115
THE YÖRÜK OF SOUTHEASTERN TURKEY 117
THE BASSERI OF SOUTHWESTERN IRAN 121
AL-MURRA OF SAUDI ARABIA 124
GENERAL OVERVIEW 127

Chapter Six
Agriculture and the Changing Village 129

LAND USE AND RURAL SETTLEMENTS 131
LAND TENURE 134
TENANCY AND ACCESS TO LAND 138
VILLAGE POLITICAL ORGANIZATION 140
TRIBALLY ORGANIZED VILLAGES 142
NONTRIBAL VILLAGES 144
HOUSEHOLD ORGANIZATION 145
RURAL TRANSFORMATION: THE CASE OF EGYPT 148
RURAL TRANSFORMATION: A CASE FROM IRAN 152
GENERAL ASSESSMENT 153

Chapter Seven
Cities and Urban Life 157

THE NATURE OF THE CITY 158
THE ISLAMIC CITY: ITS STRUCTURE AND ORGANIZATION 160
CITY LIFE IN THE TWENTIETH CENTURY 170
CAIRO: A CASE STUDY 173
SQUATTER SETTLEMENTS: A TURKISH CASE 178
PLANNED CHANGE: BAGHDAD 183
SECTARIANISM IN BEIRUT 185
GENERAL OVERVIEW 186

Chapter Eight
Sources of Social Organization:
Kinship, Marriage, and the Family 189

TERMS OF KINSHIP 190
PATRILINEAL DESCENT AND PATRONYMIC GROUPS 194
THE FAMILY AND THE HOUSEHOLD 196
MARRIAGE 198
MARRIAGE ARRANGEMENTS 201
POSTMARITAL RESIDENCE 204
CONJUGAL VERSUS DESCENT TIES 206
RESIDENTIAL PATTERNS 208

Chapter Nine
Women and the Moral Order: Identity and Change **211**
WOMEN AND THE VEIL 212
SEXUAL MODESTY 213
HONOR AND SHAME 217
SOCIALIZATION AND SEX ROLES 220
CHANGING ROLES OF WOMEN 224
LEGAL REFORM AND PERSONAL STATUS 229
WOMEN AND SELF-IMAGE 231
THE INDIVIDUAL IN A CHANGING WORLD 233

Chapter Ten
Local Organization of Power:
Leadership, Patronage, and Tribalism **241**
THE ENVIRONMENT OF POLITICAL BEHAVIOR 242
THE POLITICS OF SOCIABILITY 244
MEDIATION AND PATRONAGE 245
STYLES OF VILLAGE POLITICS 246
LOCAL-LEVEL LEADERSHIP 248
AXES OF LOCAL POWER 248
THE MORAL BASIS OF POLITICS 250
ISLAM AND LOCAL POLITICS 252
THE POLITICS OF STRATIFICATION 255
TRIBES AND TRIBALISM 257
TRIBAL STRUCTURE 259
EGALITARIANISM AND TRIBAL SOCIETY 263
TRIBALISM AND THE STATE 264
TRIBAL, PERSONAL, AND ETHNIC IDENTITIES 267

Chapter Eleven
Challenges and Dilemmas:
The Human Condition in the Middle East Today **269**
POLITICAL CENTRALIZATION 270
ECONOMIC DEVELOPMENT 274
SOCIAL AND CULTURAL DIMENSIONS OF CHANGE 278

Index **284**

PREFACE

This anthropological essay grew from more than a decade of teaching and research by both authors in various countries of the Middle East. As friends and colleagues of long standing, we have maintained an ongoing dialogue about the craft of anthropology and what it offers the student of Middle Eastern society. Our different but complementary points of departure and field experiences have helped, we feel, to make our discussions particularly fruitful. This book has no senior author. Indeed it would not be possible to attribute any passage or section to any one of us; the venture was truly collaborative in all respects. Even the actual, sometimes tedious, process of writing and revising was done in joint sessions.

As the title suggests, we see this book as an exercise in social anthropology. To that extent, our objectives are to provide a synthesis of what we feel our discipline has been able to contribute to understanding this important area of the world. We have avoided theoretical polemics and specialized jargon in the hopes of avoiding a common social-science tendency to mystify and thereby to explain less than is already known by common sense.

Our point of departure—in fact, the assumption underlying the analysis we provide—is that explanations of cultural institutions and social processes must be relatable to the behavior of individuals, their needs, values, and motivations. Individual decisions and actions, indeed behavior of all sorts, take place in the context of particular social or cultural settings. These social settings are themselves shaped by the momentum of a specific history. Not only do the material constraints of the moment affect the strategies of individuals today, but the particular ways in which material problems or opportunities were handled in the past also influence current choices.

Systems of values, norms, and religious beliefs are also an integral part of social process. Not only does an ideational system give meaning to individuals' actions, but the system is itself a source of constraints facing the individual and society as well as an arena in which people compete. Individuals and groups use ideologies of all sorts as they strive for power—to control resources, gain prestige, and influence outcomes. Although we have not paid great attention to national-level politics and economics, we have consistently kept this larger and important context in view as we developed our analysis. It would not be too much to say that a full understanding of the activities of people in the most remote village today requires an awareness of how that community fits into a national, indeed a world, economic and political system.

Many people are to be thanked for the assistance they have rendered us directly or indirectly. We cannot thank them all. Both of us were students of William Schorger at the University of Michigan at critical junctures of our intellectual growth. To him we express that special thanks due to one's teacher. We also want to acknowledge our intellectual debt to another of our teachers, Eric Wolf, whose work and thought have continued to stimulate and challenge us. In fact, at the outset of this enterprise, we took as a model his early book, *Sons of the Shaking Earth,* which describes the cultural history of the valley of Mexico. To us, this book exemplifies a superb treatment of a complex historical and cultural tradition, a treatment at once elegant, sympathethic, and honest.

Of many friends and colleagues who assisted us by their critical reading of parts of the manuscript, we would like to thank Ülkü Bates, Judith Berman, Francis Conant, Carol Kramer, Ayse Kudat, Susan Lees, Edward Malefakis, and Paul Sanfaçon. One reader deserves special acknowledgment: Gregory Johnson read the entire manuscript and offered detailed suggestions for its improvement. We also wish to acknowledge the helpful comments provided by the following individuals who reviewed the manuscript for Prentice-Hall: William Irons of Northwestern University and Charles L. Redman of State University of New York at Binghamton. Kathleen Borowik prepared all maps and illustrations, for which we are grateful. All photographs not otherwise credited were supplied by the authors.

The system of transliteration generally used is that recommended by *The International Journal of Middle Eastern Studies*. However, we have deviated from it where purposes of clarity might be served.

DANIEL BATES
AMAL RASSAM

INTRODUCTION

In this book we treat a particular region, the Middle East—its peoples and its cultural heritage—in light of what we feel to be the best available sources. Our principal objective is to impart as much information about the society and culture of this vast area as can be reasonably synthesized and interpreted in a relatively short volume. The goal we have set for ourselves is thus straightforward, even though the materials we work with are complex and refractory. Our emphasis on the synthesis of information and factual material is deliberate. Informed discussion and analysis necessarily start with a shared body of conceptualizations and facts, be they rooted in commonplace observations or in relatively esoteric knowledge. This venture, jointly undertaken by two anthropologists, is an attempt to provide materials and ideas suitable to informed discussion and debate, and thereby to provide the basis for further synthesis and analysis. Even as we introduce the student to the peoples and cultures of the Middle East, we also hope to challenge our colleagues to reassess some of their assumptions.

The term *Middle East,* however delineated, has become accepted in common usage, replacing such earlier references as the Near East, the Levant, the Holy Land, and so on. Although the term *Middle East*

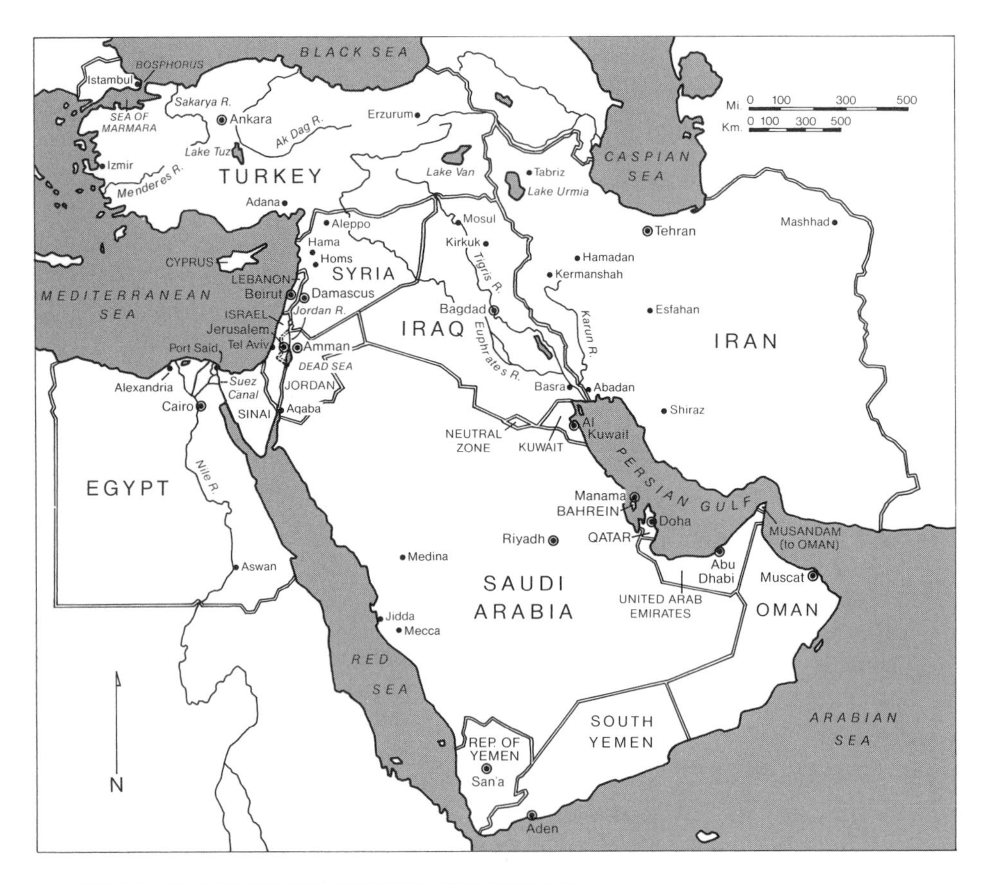

POLITICAL MAP OF THE CENTRAL MIDDLE EAST

was not employed by the peoples of the area, it has recently been accepted and used by them and, in its broadest construction, refers to a part of the world that encompasses all of the Arab states, Turkey, Iran, Israel, and Afghanistan. Without entering into the unprofitable debate of what does and does not properly constitute the Middle East, for our purposes we take the term to refer to a geographical and cultural area that includes Egypt, Israel, Lebanon, Syria, Jordan, Iraq, Saudi Arabia, the two Yemens, Bahrain, Qatar, Oman, the United Arab Emirates, Kuwait, Iran, and Turkey. We have thus excluded the nation-states of Northwest Africa, the Sudan, and Afghanistan. When compared with the wider connotation of the term, the area we are concerned with might, with justification, be called the *Central Middle East*. Even within this area we are not concerned with a country-by-country discussion;

the perspective we offer focuses on generalizable problems, patterns, and cultural processes.

The Central Middle East, as we have delineated it, acquires considerable coherence when considered from a number of perspectives. Historically this is the "cradle of civilizations," an ancient heartland of empires and cities that includes those of Egypt and Mesopotamia. Urban life and state forms of political organization first arose here. For millennia, in ever-changing configurations, great stretches of this region have been politically and economically integrated. This is also the birththplace of the three major monotheistic religions: Judaism, Christianity, and Islam. All three were shaped in the context of Middle Eastern civilization and all contribute to its ongoing expression in the lives and activities of Middle Eastern peoples.

More immediate are a number of geopolitical facts. Since the eighth century most of the great Islamic empires have had their centers located in this heartland area. In fact, the most recent of these, the Ottoman Empire, held sway over most of this region for almost 500 years, and even though their domains included much of North Africa, the latter area (except for Egypt) was peripheral to the Empire. Such political facts, together with the presence of long-established routes of coastal and internal communication and trade, impart a high level of cultural integration to the Central Middle East. Today the oil wealth found in and near the Gulf area and the large concentrations of population in Egypt, Turkey, and Iran give this central zone great strategic importance as the superpowers vie for influence and control.

One can easily argue that the French colonial experience and the presence of large Berber-speaking populations set Northwest Africa apart. Similarly, the Sudan and Afghanistan partake more of the African and Central Asian cultural experience. But these are not our main reasons for focusing our discussion on the Central Middle East. We take this narrower geographic perspective because it allows us to draw on a relatively closely interrelated body of ethnographic and historical data, thus facilitating a more systematic body of generalizations. Even with this self-imposed limitation, we feel confident that much of the interpretive discussion we offer can be useful to those interested in North Africa or Afghanistan. Much, too, of the synthesis applies to Israeli society, particularly to the Arabic-speaking population. No attempt is made here to systematically explore those cultural and political features which so sharply distinguish Israel from other Middle Eastern countries.

The Middle East, however defined, presents an almost unique challenge for anthropologists. Its long historical record, literate cultures, and established traditions of indigenous scholarship, while providing a wealth of material, make great demands on the individual scholar. Moreover, the rivalry between Christian and Moslem states for control of the Mediterranean, not to mention the more recent Western colonial ventures, add a polemical dimension to any discussion of the region. Also there is a long-established European scholarly tradition, usually called *Orientalism,* that has tended to view Islam and Moslem society as unique and esoteric phenomena, often sketched in sharp contrast to the European or Judeo-Christian experience. Although this scholarly tradition has contributed much useful insight, the Orientalist approach runs directly contrary to the basic anthropological perspective that seeks to understand variations in institutions and cultures in terms of universal human imperatives and processes. As anthropologists, we draw heavily on the work of Orientalists, while retaining what we feel to be the most useful analytic tools of our own discipline.

Although this book is meant to be an essay in social anthropology, we have not limited ourselves to the traditional concerns of that discipline. We rely heavily on the work of historians, economists, sociologists, political scientists, and geographers in order to provide a comprehensive picture of an important part of the world. We have deliberately adopted a straightforward style and nontechnical vocabulary in an effort to facilitate a discussion with as wide and varied an audience as possible.

Even though we build on the work that has gone before, the intellectual history of Western scholarship on the Middle East is beyond the scope of this book. Modern social-science research in the area really began in the 1920s, and only well after World War II do we find a body of data and theory building cumulatively on systematic and comparative research. This is not to deny the immense value of such earlier observers of Islamic society as Edward Lane, Richard Burton, Charles Doughty, and others. In fact, ethnographic reports by trained social scientists often fall short of such works as Lane's *The Manners and Customs of the Modern Egyptians* or J. L. Burckhardt's *Notes on the Bedouins and the Wahabys,* both genuine classics of their genre.

The attention of earlier anthropologists working in the Middle East was most often focused on such topics as tribal organization, patterns of marriage and kinship, and community studies of nomads and peasants. These concerns remain important. However, prior to the 1950s, little attention was paid to urban society, nontribal forms of political organization, rural-urban networks, religious institutions, and processes of social change. Today, although much is still to be learned from the close observation of small communities, anthropologists increasingly formulate their research in terms of theoretical problems and concerns that lead them beyond the local community. For the Middle East, as for other regions of the world such as Latin America and Southeast Asia, the disciplinary boundaries among social historians, political scientists, cultural geographers, and anthropologists are hard to distinguish. All share an interest in understanding the human condition as they draw on a common pool of materials and conceptualizations. All, too, are concerned with the problems of the day and the shape of the future.

At this juncture, a brief synopsis of the material to follow serves to orient the reader to how we conceive of this book in terms of the selection of topics and the rationale for their organization.

The book is organized into three general sections, each building on the preceding. The first section establishes what we feel to be the basis for any discussion of contemporary life and culture in the region. In the opening chapters we describe the geographic setting, and we sketch some of the ecological constraints and material resources on which the societies rely. These constitute the backdrop against which Middle Eastern society has to be understood. Considerable attention is then paid to Islam as an ideational structure—that is, Islam as a source of shared meaning and a system of cultural communication. As such, Islam also has to be understood as a vehicle for political action. Chapter 4 offers a general description of ethnic and communal differences, especially those of religion, sectarian affiliation, and language.

The three chapters that follow our discussion of ethnicity can be considered the second section. These constitute an ethnographic excursion treating rural and urban society and an effort to depict the ongoing social and economic processes that currently shape the social landscape. These chapters are concerned with nomadic pastoralism, village life and agricultural production, and urban society. They deal at length with how people acquire access to resources, exercise leadership, and contend for social status. The three discussions, although they share many concerns, are not meant to be entirely parallel. The discussion on pastoralism includes substantial sections dealing with the life and social or political organization of particular peoples. This is, we hope, a means of introducing the human aspect of ethnographic analysis. The chapter on village life and agriculture contains a lengthy analysis of changing patterns of land use and settlement as they pertain to the ongoing transformation of the Middle Eastern countryside as a result of commercial farming and mechanization. The chapter dealing with urban society is directed, in some measure, toward an understanding of the current patterns of urban growth as well as the problems associated with it. The objective throughout is not to present comprehensive ethnographic reviews of the literature, but to depict the sources of continuity and the sources for change that are at work in all sectors of Middle Eastern society.

The third area around which we have organized our inquiry addresses a number of issues which, while least amenable to generalization and synthesis, are central to understanding the broad texture of Middle Eastern society and local politics. We pay considerable attention to what some have called *primordial systems* of social and political organization and interaction or, more simply put, *kinship*. We try to see the individual in Middle Eastern society as an actor within an institutional and cultural context in which the actor's behavior unfolds. We treat changing sources of personal identity, changing gender roles, and the principles by which social groupings are constituted and transformed. We examine the nature of political leadership and the local organization of what, in essence, are relationships of power and contention. In this analysis, the tribal idiom of political action is viewed in terms of other, alternative vehicles of recruitment and mobilization, such as religion, the state, or even individual charisma. We conclude the book with a tentative sketch of what we regard to be the major transforming processes underway, their economic and social sources, and some of their consequences for national integration and development.

CHAPTER ONE: THE SETTING: HUMAN GEOGRAPHY AND HISTORICAL BACKGROUND

The sources of water are the sources of life itself in the Middle East. The distribution of people, the settlements they have created, and the ways they secure their livelihoods are closely shaped by the challenges of securing and controlling this vital resource. To be sure, access to water is only one among many environmental problems faced by the inhabitants of the region. Other challenges include crop-threatening and life-threatening diseases, poor soil quality, extremes of temperature, and ever-diminishing sources of ligneous fuel. However, one need only compare a rainfall map of Europe with one of the Middle East to appreciate the magnitude of the problem posed by water. Indeed, geographers classify most regions of the Middle East as being rain deficient.[1]

Water, whether its source be rainfall, rivers, or subterranean aquifers, is the primary limiting factor governing human habitation. Beyond the frontiers of cultivation lies desolation, and the towns and villages of the Middle East closely hug the rain-swept coasts and mountain valleys, the courses of major streams, and scattered oases. As the archeologist Robert Adams writes with respect to the ancient heartland

[1] See, for example, W. B. Fisher, *The Middle East: A Physical, Social and Regional Geography*, 5th ed. (London: Methuen, 1963).

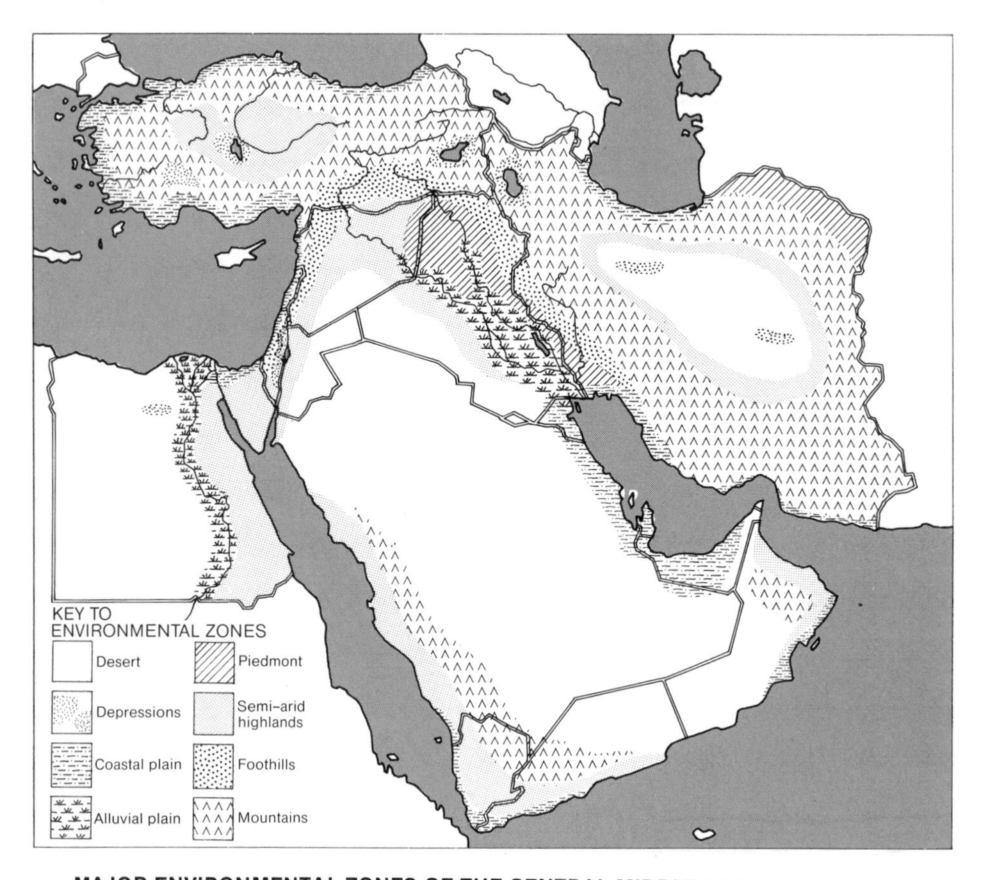

MAJOR ENVIRONMENTAL ZONES OF THE CENTRAL MIDDLE EAST

of Mesopotamia, the prevailing uncertainties about water underlie all the human adaptations, be they farming or pastoral.[2]

It is thus impossible to separate cultural or social processes from the environmental setting in which they occur. The geological or topographical configuration, the climatic conditions, the distribution of water and minerals, and the occurrence of plant and animal life all affect the way in which people live, and indeed how society has developed. Of course this is a two-way process, as human activity itself has had a profound impact on the natural habitat. The physical environment

[2] See Robert McC. Adams, *Land Behind Baghdad: A History of Settlement on the Diyala Plains* (Chicago and London: University of Chicago Press, 1965); and Robert McC. Adams, *Heartland of Cities: Surveys of Ancient Settlement and Land Use on the Central Floodplain of the Euphrates* (Chicago and London: University of Chicago Press, 1981); and Robert McC. Adams and Hans J. Nissen, *The Uruk Countryside: The Natural Setting of Urban Societies* (Chicago and London: University of Chicago Press, 1972).

and the landscape we describe for the Middle East are at once the setting for and the result of a long history of human adaptation. The face of the Middle East we see today is as much history as it is geography.

THE GEOLOGICAL FOUNDATIONS

Carleton Coon, an anthropologist who has worked extensively in the area, notes that, "Geologically speaking, the world's oldest civilizations arose in some of the world's youngest lands."[3] The geological youth of the Middle East is evident in the contrast of the steeply rising, craggy mountain ranges which form an often spectacular backdrop to the flat plains and steppes. The process of recent mountain formation has determined the distribution of critical resources in the area, including the sources of water and minerals. The mountains of the Middle East consist of sea-deposited layers of sandstone and limestone. Consequently, these are, apart from earth itself, the most commonly used building materials in the region today, and they were used to produce some of the best-preserved monumental architecture of antiquity. However, these recent deposits of sandstone and limestone do not contain metallic ores, and it is only where mountains have folded and faulted to expose underlying older layers that iron, copper, tin, silver, and other minerals are found. These minerals, however, are restricted to relatively few sites; the most important historically have been the iron and copper mines of eastern Anatolia, the copper of the Zagros and Sinai, and the silver of the Taurus. Few minerals are produced in exportable quantity today; indeed, with the important exceptions of oil and bauxite, the Middle East imports its industrial-grade minerals from abroad.[4]

Regional and local topography are important in determining where people settle, how cities develop, the arrangement of trade and trade routes, and even political organization. For instance, the Nile Valley by virtue of its topography and river-facilitated communications, lends itself to unified political control, whereas the rugged, cross-cutting mountains of the Levant do not. Topography further directly influences local climatic regimes, which in turn determine the potential for human and animal life.

If you were to fly over the area, your immediate impression would be of a landscape alternating between rugged mountains or plateaus and lowland areas where the boundaries between the desert and the sown land are often starkly etched. This impression is fairly accurate in that it

[3] Carleton Coon, *Caravan: The Story of the Middle East* (New York: Holt, Rinehart & Winston, 1958), p. 10.

[4] See Fisher, *Middle East*.

reflects the way in which most geologists and geographers classify and describe the landforms in the area.

The geology of the area is often described as being made up of two rather different landform systems. On the one hand, the vast lowland areas that make up the bulk of the Arabian Peninsula and Egypt are part of a large, geologically stable, and for the most part highly arid shield or massif that extends from Africa to India. In contrast, the second landform system to the east and north of the Arabian shield is a geologically active mountain zone subject to recurrent earthquakes. This encompasses the great Iranian and Anatolian plateaus. Lying between the Iranian plateau and the Arabian shield, the broad plains and marshes of Mesopotamia form a continuous depression subject to massive alluviations from the rivers that traverse it.

The unity of the first system, which Cressey suggests may have formed one ancient continent, is interrupted by the recent faulting that created the Dead Sea, whose floor is 395 meters below sea level.[5] This broad area is one where horizontal layers of sedimentary deposits lay largely undisturbed, and where hills rarely rise above 1000 meters. One place in which we can see the ancient sedimentary deposits exposed is in Egypt, where the force of the river Nile, flowing from its headwaters in tropical Africa to the Mediterranean Sea, has laid bare the various layers. Few visitors to Luxor, the ancient capital of Upper Egypt, ever forget the effect of the day's last light playing on the red-hued cliffs into which the Pharaohs carved their magnificent tombs. Egypt, which has been called "The Gift of the Nile," has its productive heartland in this valley, which the Nile has carved out of the desert waste.

To the east, the great Arabian shield is bounded by a major riverine system, that of the Tigris and Euphrates. Here, as in Egypt, we find the heartland of ancient civilizations. In contrast with Egypt, however, the Tigris and Euphrates valley is broad, with regular shifts in the rivers' courses occurring each spring as the rivers follow a wide geologic depression to the Persian Gulf from their headwaters in the mountains of the north and east. The Euphrates is, again in Adams's words, "a brown, sinuous pulsing artery that carries the gift of life."[6]

Elsewhere, it is the sedimentary layers of the Arabian massif formed by ancient seas that harbor the petroleum wealth of the region. In 1978 it was estimated that over 62 percent of the world's proven oil reserves lay in the Middle East, including North Africa, and that 42 percent of all oil produced originated in this part of the world. The proven reserves of Saudi Arabia alone are a staggering 25 percent of the world's

[5] G. B. Cressey, *Crossroads: Land and Life in Southwest Asia* (New York: Lippincott, 1960), p. 68.

[6] Adams, *Heartland of Cities*, p. 1.

total. Iran, Iraq, and Kuwait jointly account for another 25 percent, with the result that this relatively small geographic region, focused on the Arabian Peninsula and the Persian Gulf, dominates the world's energy market. Geologists now speculate that Iraq has yet undiscovered or unpublicized reserves which may ultimately rival those of Saudi Arabia.[7] On the other hand, before the revolution slowed production, Iran was expected to exhaust its proven reserves within a period of 30 years or so.

The second major landform system comprises an extensive mountain zone, part of a large one that reaches from the Alps to the Himalayas. Perhaps the best way to picture this complex of craggy and often disconnected local ranges is to visualize two roughly parallel zones of transcontinental foldings. One stretches from the northern Alps through northern Turkey, Iran, and Afghanistan, and culminates in the Himalayas. The second zone, rising with the southern Alps, extends along the southern Turkish coast, northern Iraq, and Iran, and it also culminates in the Himalayan ranges. The northern rim of mountains in Turkey includes the Pontic range, which runs along the Black Sea and continues in Iran as the Elburz. In eastern Turkey, the highest peak is that of Mount Ararat of Biblical fame (elevation 16,946 feet). In Iran the Elburz continues south of the Caspian Sea, with Demavand as its highest peak (elevation 18,934 feet). Demavand is the highest mountain west of the Hindu Kush, and like most of the other mountains in the area, it is volcanic but dormant.

The southern mountain system in Turkey is represented by the twin ranges of the Taurus and the Anti-Taurus. Rising in the west where the Aegean meets the Mediterranean, they continue eastward, dominating the southern coastline of Turkey and leaving a relatively narrow coastal strip. The Taurus range is relatively unbroken until it reaches the northeastern extremity of the Mediterranean where the hitherto narrow coastal strip opens into an extensive plain, the Amik. Broken by a major pass, the Cilician Gates, the range then continues as the Anti-Taurus in southeast Turkey and the Zagros in Iraq and Iran. Both the northern Pontic-Elburz and the southern Taurus-Zagros ranges are punctuated by many alluvial valleys, frequently rich in soil and relatively well-watered, although separated from one another by terrain difficult to traverse.

The major mountain systems in Turkey and Iran enclose and isolate two large interior plateaus comprising most of the land mass of this geological zone. The Anatolian and Persian plateaus vary in elevation from 1650 feet to 4950 feet and are bordered by ranges whose peaks

[7] James A. Bill and Carl Leiden, *The Middle East: Politics and Power* (Boston: Allyn & Bacon, 1974), pp. 364–65.

frequently rise above 6600 feet. Rimmed by high mountains, these inland plateaus are characterized by internal drainage resulting in a number of brackish or alkaline lakes, the most outstanding of which are Lake Van in Turkey and Urmia in Iran. Of greater economic import may be the fact that areas of central Turkey and still greater areas in Iran consist of salt flats or virtual deserts such as the Dasht-i-Kavir and Dasht-i-Lut of Iran.

Although it is perhaps possible to sketch the broad geographical outlines of the Middle East in terms of these two contrasting geological systems, the picture is far from complete, as it ignores the most critical elements for human activity—namely, the distribution of surface water and vegetation. From this perspective, the interface between the just-described geological systems constitutes a major area in itself, one that stretches from the Persian Gulf northward and westward along the Mediterranean and ends in the delta of northern or lower Egypt. This encompasses most of the areas of classical civilization—Mesopotamia, the Levant, and Egypt—all sites of great agrarian societies in antiquity. This is the original area of sown land, where it seems that the earliest human experimentation with the domestication of plants and animals took place.

CLIMATIC REGIMES AND WATER

Within this area of interface lies the so-called Fertile Crescent, a grassland steppe, itself a result of the annual regime of wind-borne rains caused by winter westerlies from the Mediterranean. However, no community in this area is far removed from nearby zones of considerable aridity. It is this marked local contrast between areas of cultivation and rangelands or deserts that is one of the most striking environmental characteristics of the Middle East. This contrast is largely the result of the effect of the mountains on the moisture-bearing winds that seasonally move inland, dropping the bulk of their water on the seaward-facing slopes. In fact, only Lebanon is without significant areas of extreme aridity.

In the southern part of the Arabian Peninsula, the summer monsoons coming across the Indian Ocean strike the green mountain, Jabal Akhdar of the Hadramaut, and the mountains of Yemen to make parts of the latter some of the most fertile areas of the peninsula—the fabled *Arabia Felix*. In the past high rainfall here permitted local populations to grow fine coffee, an endeavor largely abandoned today. The spread of coffee as an international beverage can be traced back to its Yemeni origins and its export from the coastal town of Mocca. Immediately behind this monsoon-drenched, sea-facing escarpment, however, begins

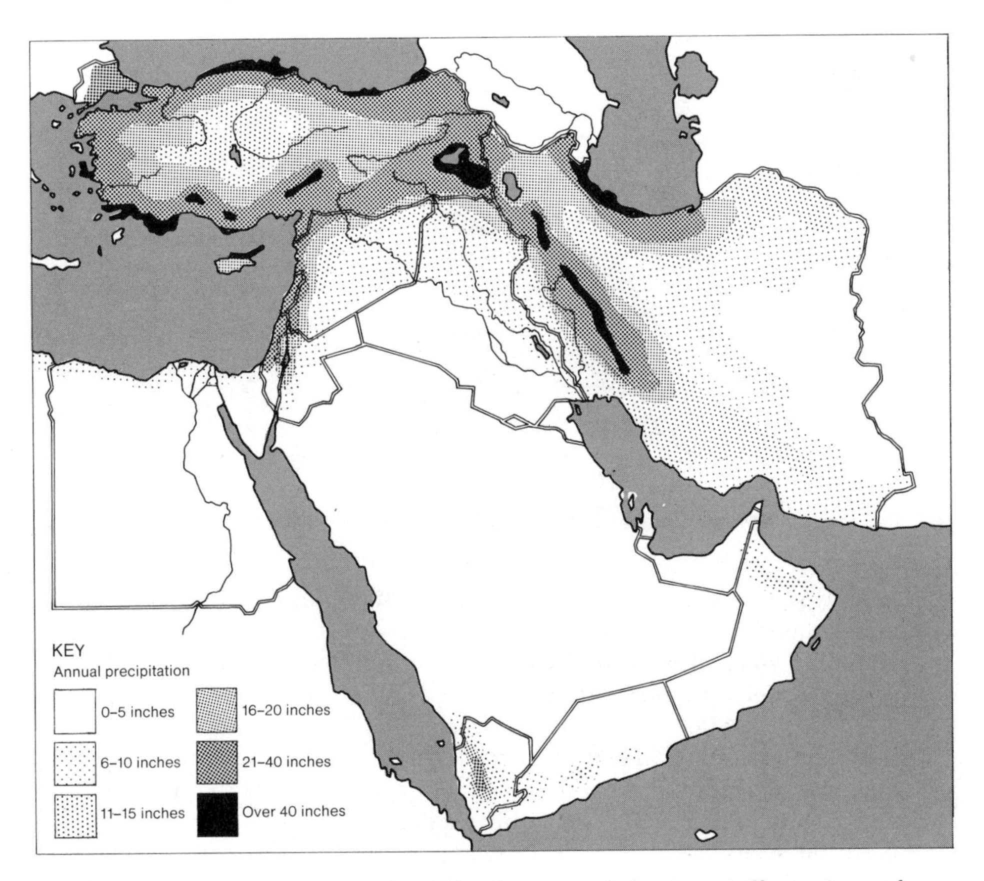

RAINFALL MAP Rainfall together with other sources of water sets the contours of agriculture and the distribution of settlements.

Arabia Deserta, the region that envelops the Empty Quarter, or Rub'al-Khali, a vast sand desert largely devoid of human settlements.

In most of the Middle East, westerly rain-bearing winds coming from the Atlantic and across the Mediterranean sweep inland during the winter months. From the point of view of water loss in this region of high summer temperatures, winter rains arrive at the ideal time. Less water is lost through evaporation—always a problem in this arid zone—and many indigenous plants, such as barley, wheat, lentils, and chickpeas, are adapted to a winter growing season. Even so, conditions for plant growth are highly constrained because the winter cold impedes maturation, and the spring season of maximal growth is short. These westerlies are, moreover, extremely capricious, so that the amount and distribution of annual rainfall is not predictable. Villages experiencing severe drought one year may be contending with floods the next. As we have

Nomads of the Egyptian eastern desert drawing water from desert well. (Photo courtesy of Diana de Treville)

already seen in the case of southern Arabia, mountain topography determines the strikingly uneven distribution of moisture. The spatial and temporal variability in both rains and river flooding constitute fundamental uncertainties with which people must cope.[8]

The often lush, well-watered western slopes and plateau edges of the mountains stand in marked contrast to the aridity and desiccation of the eastern-facing slopes and valleys. For example, the Sinai Peninsula and eastern Jordan, which lie within such a rain shadow, experience virtually no rainfall at all. On the other hand, the coastal strip of the Levant or eastern Mediterranean coast, with an annual rainfall of over 30 inches per year, receives more rain than some parts of the British Isles. However, Damascus, an hour's drive inland through the Lebanon range, receives a mean annual rainfall of less than 10 inches. The Mediterranean and Aegean coastal plains of Turkey exhibit a similar pattern. Along the coast, rich soils support lush groves of orange trees, cotton, and even bananas in one district—that of Alanya. However, a short drive by car over the Taurus brings one to the Anatolian Plateau, where rainfall agriculture is limited to drought-resistant grain crops and legumes, and is often precluded altogether due to either aridity or soil salinity.

Although 10 inches of annual rainfall is considered to be the minimal requirement for dry farming of wheat, estimates of average rainfall in the area are virtually meaningless. For example, in the Da-

[8] See Adams, *Land Behind Baghdad;* Adams, *Heartland of Cities;* and Adams and Nissen, *The Uruk Countryside.*

mascus region, rainfall agriculture is impossible because the rains vary greatly from year to year and much of the rainfall may come in the form of one or two cloudbursts. Thus, in the interior of the Middle East, wherever agriculture depends on rainfall, it becomes a risky venture. Virtually the only exceptions to this are such restricted areas as the Black Sea coast of Turkey, the Caspian coast of Iran, and parts of highland Yemen. Most of Arabia and Egypt are rainless and, in fact, not more than 10 percent of the total land mass of the Middle East is suitable for cultivation for this reason. The most pervasive and enduring strategy of land use in areas where rainfall agriculture is feasible is to pursue very extensive forms of field use, with long fallows and a variety of crops planted, in hopes that, despite poor rains, some will last until harvest. Cultivation is often combined with animal husbandry to further diversify the household's resources in the face of high risk. Animals are more than an alternative form of food production; they are also a means of storing food on the hoof.

Although the distribution of peoples and settlements in the Middle East closely reflects the distribution of rainfall, this does not complete the picture of sown areas. Although agriculture depends on regular water supplies, these can be obtained in ways other than by rainfall. First perfected in the Middle East, irrigation agriculture was a major technological advance, which allowed for the development of a very complex social order. The two civilizations of classical antiquity—Egypt and Mesopotamia—owed their florescence to the irrigation systems developed along the river valleys of the Nile and the Tigris-Euphrates. Of the Euphrates, Adams writes that despite the vicissitudes of shifting courses, unpredictable annual flow, and periodically devastating floods, "It has provided the only possible foundation for an immense column of human achievement that has risen laboriously in a pivotal region over hundreds of generations."[9] Precisely how the development of irrigation influenced the course of early civilization is a complex question but, without doubt, even early efforts at water control increased agricultural productivity and ultimately generated the food surpluses to sustain segments of the population not directly involved in cultivation.

Water control and management in general are ancient sciences in all of the area. As water is life, the people of the Middle East have ingeniously devised many and varied means of conserving it and extending its distribution. Water control in the form of dams and irrigation schemes is still undertaken as major public-works projects by governments of the region. The massive Aswan Dam of Egypt is only one of many; others are the large Keban project of eastern Turkey, the

[9] Adams, *Heartland of Cities*, p. 1.

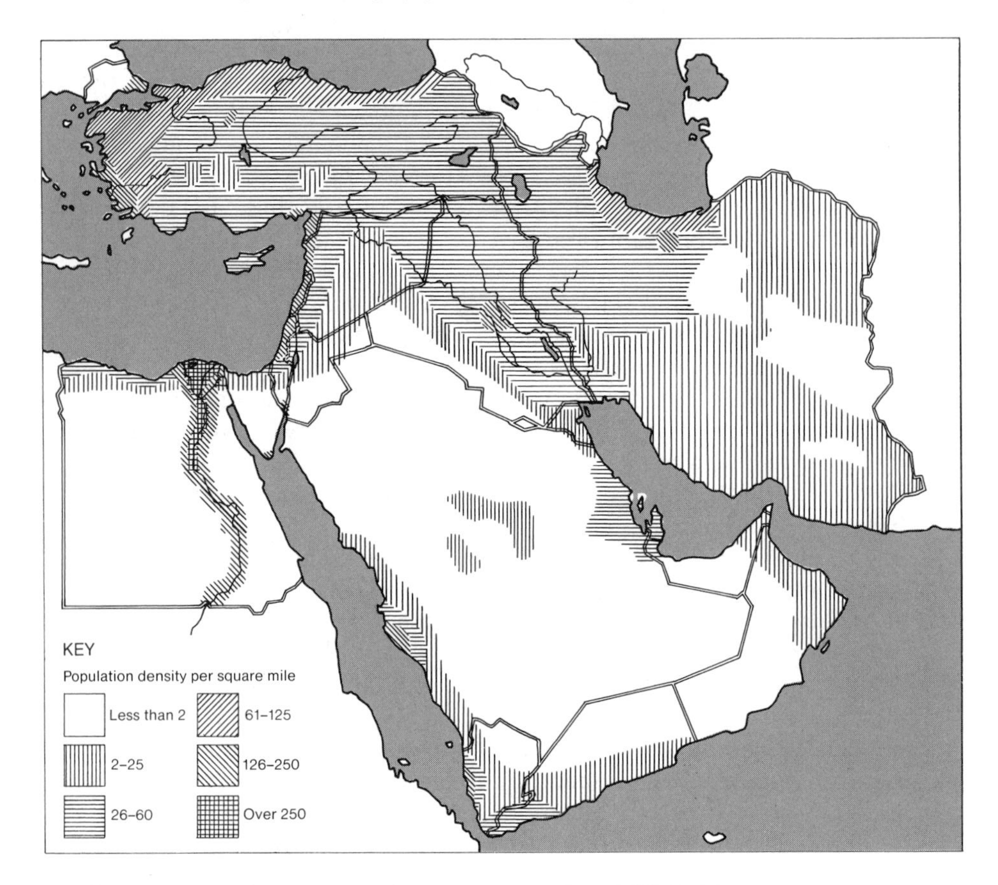

DISTRIBUTION OF POPULATION IN THE CENTRAL MIDDLE EAST

Euphrates Dam of Syria, and the Kur river project of south-central Iran. All of these further illustrate that water is the single most important factor limiting agriculture in the Middle East.

Although the politics of large-scale hydraulic projects attract the most attention, of far greater importance to the peoples of the arid zones of the Middle East are the highly varied local techniques for water management. The *qanats* of Iran and the *aflājs* of south Arabia are underground tunnels that carry water from upland sources and thus create artificial oases and extend village cultivation into the desert.[10] An entire string of small communities may come to depend on one water source, and households may share water rights and responsibilities for its maintenance.

[10] J. C. Wilkenson, *Water and Tribal Settlement in South-East Arabia: A Study of the Aflāj of Oman* (Oxford: Clarendon Press, 1977).

The ancient Egyptians developed three devices for lifting water from canals and basins to their fields: the *shadduf,* or weighted pole with a bucket on the end, the Archimedean screw, and the water wheel powered by animal traction. These techniques, few among many, were early achievements of the agricultural peoples of the region. Increasingly, traditional methods of securing and lifting water were replaced by tube wells and motor-driven pumps.

Although techniques of irrigation have improved dramatically, sometimes opening new areas for cultivation, one age-old problem remains—that of soil salinity.[11] Wherever the water table in this area of high evapo-transpiration rises to about 1.5 meters from the surface, salts occurring naturally in the soil are drawn upwards. This leaching of salts ultimately contaminates the soil to the point of diminishing yields, and may even preclude planting altogether. In Iraq and elsewhere, this problem has led to the abandonment of much otherwise tillable acreage for which reclamation would be a costly enterprise. The use of modern pumps often exacerbates the problem of salinity because it encourages overirrigation, particularly where river water is available. Where tube wells are used, another frequently encountered problem is the depletion of stored water reserves, which are not recharged by rains in this rain-deficient region.

Another factor affecting human life in the Middle East is temperature. The Middle East generally experiences hot, dry summers and cool, wet winters, making much of it generally Mediterranean in climatic regime. For example, the average winter temperature in Tehran (elevation 4000 feet) is 36 °F, and the daily summer average for three months is 86 °F. Some cities in arid, low-lying zones experience consistently higher temperatures throughout the year. In Baghdad, the captial of Iraq, the temperature may occasionally exceed 100 °F during a seven-month period, with a July extreme of up to 121 °F.[12] Furthermore, in most parts of the Middle East summer temperatures are exacerbated by hot, dust-laden winds known variously as the *sirocco, sharqi,* or *khamsin.* These winds, blowing from the south and southeast, often reach gale force and contribute to the problem of desiccation by removing the top soil layers.

The climatic regime in the Middle East, with its extremes in temperature and precipitation everywhere, requires that local inhabitants invest heavily in shelter. In traditional Iraqi homes, substantial basements offer cool daytime summer refuges for members of the household, while the flat rooftops provide much-prized relief during the nights. In the upland regions of the Anatolian and Persian plateaus,

[11] This was a problem even in antiquity. See Thorkild Jacobsen and Robert McC. Adams, "Salt and Silt in Ancient Mesopotamian Agriculture," *Science,* 128 (1958), 1251–58.

[12] *Oxford Regional Economic Atlas* (Cambridge, Eng.: Cambridge University Press, 1979).

winter blizzards of great severity are not unknown, and mountain passes are often closed by snow. Each winter and early spring in the mountains of Iran, Turkey, and Iraq, thousands of villagers are temporarily cut off from the world by heavy and long-lasting snowfalls and their melt water.

Landforms, climate, and water combine to establish the distribution of natural vegetation. Even slight variations in altitude, precipitation, or range of temperature have great consequences for plant life and, by extension, for food production and even for the availability of fuel and building materials. Wood fuel and lumber are restricted today to relatively constricted high-level areas. However, these are receding rapidly in the face of pervasive overgrazing and heavy exploitation for household use. As the forest and brush cover diminishes, ever-increasing amounts of topsoil are carried off by the winter rains or spring melting of snows, which further limits the propagation of most tree species.

Along the Black Sea coast of Turkey and the southern shores of the Caspian, we see the last remnants of formerly extensive deciduous forests. In the Taurus range, parts of the Zagros, and sporadically elsewhere, stands of conifers are found where enough moisture is available. Like the deciduous forests, they too are in retreat in the face of great demand for firewood, charcoal, and building material. Elsewhere, low-growing shrubs are the predominant growth. In the much more extensive nonforest areas of the highlands, open areas support Alpine or sub-Alpine grasses, depending on altitude and rainfall. The vast reaches of Arabia, the Syrian steppes, and the more arid portions of the Anatolian and Persian plateaus are characterized by rugged plants that take advantage of brief irregular rains by rapid growth and bloom, followed by long periods of dormancy. In the arid areas, trees are generally found standing as a ring of green sentinels around village settlements only where tended by humans for building or fuel purposes. Desert and steppe grazing cycles were formerly determined by the availability of water for the flocks. Today, with mechanized transport of animals and water, grazing pressures are far heavier than before. Overgrazing is causing the rapid reduction of desert flora, including brush cover in southern Arabia, Syria, and Jordan.

POPULATION AND SETTLEMENT PATTERNS

Given the diversity of landform and climate, population is very unevenly distributed in the Middle East. Overall, the region remains one of the least densely populated in the world; some areas are virtually uninhabited, such as the desert depressions of Iran, the Rub'al Khali of Arabia, and the Saharan deserts of Egypt, Libya, and the Sudan. In contrast, as we have noted, the well-watered alluvial river valleys are characterized by high population densities, with the Nile Valley and its delta by far the most heavily peopled area. According to a 1978 estimate that includes urban areas, it has a density of 1200 to 1400 people per square kilometer of arable land.[13] Thus, 99 percent of Egypt's population is concentrated on 4.5 percent of its territory!

Because most Middle Eastern countries possess large tracts of arid, uncultivable land, the ratio of agricultural population to arable land is thought to be a better index of density.[14] Such figures contrast with standard density measures (population to total area) markedly. In the case of Saudi Arabia, measuring the ratio of agricultural population to arable land for 1970 gives us a density of 395.5 people per square kilometer, whereas the *overall* density ratio is 3.6 per square kilometer—a 100-fold difference. The discrepancy would be even greater if we included the nonagricultural population per square kilometer of cultivable land, which is perhaps an even more accurate index of density. Even in the desert, oases are densely populated by cultivators and by petroleum workers as well. More than half the population of Jordan and Iraq inhabit 14 to 16 percent of their respective land areas. Even in Turkey, which has a more evenly distributed population than the other countries, regionally calculated rural densities range from 7 to 127 people per square kilometer.[15]

For the most part, population distribution is conditioned by the availability of water, though economic, historical, and political factors need to be examined to adequately explain why certain areas—for example, the lower Nile Valley, the hill country of northern Lebanon, northwest Jordan, the uplands of Syria to the east of the Orontes River and extending into south-central Asia Minor, and the uplands of Yemen—are all more densely populated than neighboring areas characterized by similar ecological conditions.[16]

The sharp contrasts in local environments and ways of life that distinguish the Middle East and its human geography can be seen in terms of the diverse challenges and problems to which people have responded in different ways. Today, we see nomadic herders establishing a succession of isolated camps through the deserts and steppes in pursuit of pasture, while in the Nile delta, Egyptian peasants concentrate in large villages of as many as 6000 people each. Meanwhile, along the slopes of the Zagros villagers eke out a living from small plots salvaged at great

[13] Fisher, *The Middle East*, p. 270.

[14] Peter Beaumont et al., *The Middle East: A Geographical Study* (London and New York: John Wiley, 1976), p. 184.

[15] Ibid, p. 184.

[16] Fisher, *The Middle East*, p. 272.

Terraced field in North Yemen.

expense of labor. These patterns, which may impress us today as timeless, have their origins in particular times and places. In other words, they have their history. And to understand the Middle East today, we must know something of this history, since human societies are shaped not simply by their responses to problems of the present, but equally in terms of how they solved those of the past.

PREHISTORIC PATTERNS OF ADAPTATION

Although humans have lived in the Middle East for tens of thousands of years, history in one sense begins with the period known as the Neolithic, or New Stone Age. The Neolithic, which roughly dates from about 10,000 years ago, is often regarded as a watershed in the development of human culture. It was during this era that domestication of plants and animals took place, thus setting in motion profound changes in human society. In fact, some archeologists have characterized this development as a "revolution," analogous to the Industrial Revolution that so dramatically transformed the world in the last two centuries.[17] Other archeolo-

[17] See V. Gordon Childe, *Man Makes Himself* (New York: New American Library, 1951); Robert Braidwood, "The Agricultural Revolution," *Scientific American*, 203 (September 1960), 130–41.

gists, cognizant that the events of the Neolithic unfolded over a period of several thousand years, avoid the term "revolution," while still acknowledging that the advent of food production, as opposed to hunting and gathering, established the preconditions for what we usually term *civilization.*[18]

Domesticated grains, such as wheat and barley, and animals, such as sheep and goats, ultimately gave humans access to increased sources of food energy per unit of land. Perhaps of greater importance, as food production became more reliable, it allowed large numbers of people to live in areas hitherto unsuited for year-round habitation. No longer dependent on often widely scattered wild plants and game, the human population in the Central Middle East increased. Village life rapidly emerged as a more regular pattern, a prelude for the soon-to-follow development of cities and states. In this sense the Neolithic marks the point at which the Middle East culture as we know it today began to take shape. Let us briefly examine these developments, which laid the basis for adaptations persisting today.

Anthropologists recognize that people do not usually "discover" something as complex as agriculture; instead it has to be regarded as the culmination of a long series of interrelated events, even accidents. People slowly, and often without realizing it, react to specific problems in ways that only later will be seen as significant. The question we have to ask is why people in the Middle East changed their mode of subsistence to emphasize agriculture and domesticated animals.[19]

Early students of this problem, such as the archeologist V. Gordon Childe, attributed the change to environmental pressures—namely, the increased desiccation that was thought to have occurred around 10,000 years ago.[20] Childe suggests that this forced people and animals to congregate around a relatively small number of oases and river valleys, which placed them in close symbiotic relationship with each other as well as with certain plant species. Proximity encouraged prople to observe and experiment with their environment, which led in time to the taming of animals and the domestication of plants. The resulting increase in population and cultural complexity ultimately gave rise to

[18] See, for example, Kent V. Flannery, "The Ecology of Early Food Production in Mesopotamia," *Science,* 147 (1965), 1247–56; and Kent V. Flannery, "The Origins and Ecological Effects of Early Domestication in Iran and the Near East," in *The Domestication and Exploitation of Plants and Animals,* eds. Peter J. Ucho and G. W. Dimbleby (Chicago: Aldine, 1969), pp. 73–100. For a survey of ideas on the origins of agriculture in the Middle East, see Gary Wright, "Origins of Food Production in Southwestern Asia: A Survey of Ideas," *Current Anthropology,* 12, nos. 4–5 (1971), 447–77.

[19] See Flannery, "The Ecology of Early Food Production," Flannery, "Origins and Ecological Effects," and Kent V. Flannery, "The Origins of Agriculture," *Annual Review of Anthropology,* 2 (1973), 271–310.

[20] Childe, *Man Makes Himself.*

the great cities and empires of early antiquity. Childe's explanation for the cultural breakthrough that we call the Neolithic is thus predicated on the occurrence of a sudden and severe climatic change.

More recent archeological work, most notably by Robert Braidwood, Kent Flannery, and Frank Hole, does not support Childe's theory.[21] As they point out, there is no real evidence for severe climatic change at this period, nor any reason to think that such a change would necessarily have led people to domesticate plants and animals. They point out that the natural habitat for the wild ancestors of wheat and barley is not in the lowlands, but seems to lie in the higher areas of the Zagros, which suggests that the early experimentation with domestication took place far from the centers of early civilization that followed in the lowlands. In fact, even today wild barley and wheat can still be found in the Zagros uplands. Jack Harlan, a botanist, demonstrated in 1967 the great abundance and productivity of wild wheat in southeastern Turkey. Using a primitive stone sickle, he hand-harvested 6 pounds of wild wheat in an hour and estimated that a family of four could gather a year's supply of food in approximately 3 weeks.[22]

Early foragers in the area must have found such wild grains a good source of food, and it is likely that a number of pre-Neolithic foraging populations came to depend on them as their staple. Recent archeological evidence indicates a long preagricultural tradition of village life based on wild grain and animals.[23] At the same time, it is likely that not all populations had equal access to these naturally abundant grain areas. Some must have been living in areas with limited or erratic food supply. Kent Flannery suggests that, in fact, it was among these marginally located populations that early experimentation in domestication is more likely to have occurred.[24] Because domesticated grain represents a genetic change from the original form, it is possible that the pressures that precipitated this change were inadvertently engineered by humans attempting to utilize grains where they normally did not grow. For example, Hole, Flannery, and Neely describe the planting of

[21] Robert Braidwood and Bruce Howe, *Prehistoric Investigations in Iraqi Kurdistan*, Studies in Ancient Oriental Civilization No. 31 (Chicago: University of Chicago Press, 1960); Robert Braidwood, "The Agricultural Revolution," in *Prehistoric Men*, 5th ed. (Glenview, Ill.: Scott, Foresman, 1975). See also Frank Hole, Kent V. Flannery, and James A. Neely, *Prehistory and Human Ecology of the Deh Luran Plain*, Memoirs of the Museum of Anthropology, University of Michigan, Anthropological Papers No. 1 (Ann Arbor: University of Michigan, 1969).

[22] Jack Harlan, "A Wild Wheat Harvest in Turkey," *Archaeology*, 20, no. 3 (1967), 197–201; see also Jack Harlan and Daniel Zohary, "Distribution of Wild Wheat and Barley," *Science*, 153, (1966), 1074–80.

[23] Charles L. Redman, *The Rise of Civilization: From Early Farmers to Urban Society in the Ancient Near East* (San Francisco: W. H. Freeman, 1978).

[24] Flannery, "The Origins and Ecological Effects of Early Domestication."

wheat and barley close to edges of swamps at Ali Kosh, an archeological site in southwestern Iran.[25] Because these grains do not normally grow wild in marshy areas, the native vegetation must have been cleared away to make room for them. Here, suggests Flannery, we see one of the first steps in the deliberate modification of the Middle East landscape by human hands. Early efforts such as these must have ultimately led to a full-time commitment to agriculture.[26]

Even though our understanding of all the events leading to the domestication of plants and animals in the area is still very limited, it is certain that by 7000 B.C. villages based on domestic plants and animals were becoming increasingly common. Evidence from this early period suggests that early agriculturalists practiced dry farming and utilized domesticated sheep and goats. Certainly by 6000 B.C. village life replaced nomadic foraging as the dominant pattern throughout most of the Middle East. Jarmo, a site in northeastern Iraq excavated by Robert Braidwood, was a village of approximately twenty-five mud-walled houses, each with its own courtyard, storage pit, and oven.[27] Sickle blades and grinding stones, found together with barley and wheat, indicate an ongoing commitment to agriculture. Finds elsewhere, for example in Turkey, Iran, Syria, Israel, and Jordan, suggest that the case of Jarmo is not unique.[28]

Although a modern visitor to ancient Jarmo would find its inhabitants unlike any living group in the Middle East today, certain features of their general life would be familiar. The people of Jarmo, like many contemporary rural Middle Easterners, practiced a mixed economy combining grain production with animal husbandry. In an environment of unpredictable climatic variability, it is advantageous for people to hedge their bets by diversifying their subsistence base. Moreover, the presence of individual domiciles and granaries in Jarmo suggests another similarity to modern farmers in that the household, then as now, was an extremely important social and economic unit. Everywhere in the rural Middle East the household is still the primary unit of production and consumption—a point we take up later.

Although dry farming, as evidenced in sites like Jarmo in Iraq and Jericho in Jordan (the latter is perhaps the oldest continually inhabited site in the world), became rapidly established, it did not solve all the problems of subsistence.[29] As people came to rely heavily on cultivation, they placed themselves in an increasingly vulnerable position. Variation

[25] Hole, Flannery, and Neely, *Prehistory and Human Ecology of the Deh Luran Plain.*

[26] Flannery, "The Origins of Agriculture," p. 282.

[27] Braidwood and Howe, *Prehistoric Investigations in Iraqi Kurdistan.*

[28] See Redman, *The Rise of Civilization,* for discussion.

[29] Kathleen Kenyon, *Archaeology in the Holy Land,* 3rd ed. (New York: Praeger, 1970).

in rainfall, for example, usually affects farmers more severely than it does hunters and gatherers. Foragers exploit a very broad range of food sources and may find it easy to disperse and congregate according to local conditions. Village-dwelling farmers, on the other hand, invest labor in the land they till, and in their houses and tools, and they are very dependent on localized solutions to problems of food production. They find it harder to simply pack up and move or change their basic pattern of procurement.

It is therefore not surprising that efforts were made to control the critical variable for agriculture—namely, water.[30] These attempts probably involved the planting of crops along the courses of shallow rivers or in seasonally flooded valley bottoms. Eventually means were perfected to further control water with canalization. As this involved great commitment of human labor, we can assume that it arose more from necessity than from an attempt to increase production, because people, then as now, are more apt to expend labor in efforts to maintain a system rather than to alter it.

In the Middle East techniques of water control became elaborated in the arid lowlands where irrigation farming is the key to stable settled life. It is only with irrigation and the development of aridity-resistant crops that the lowlands became settled in the pattern we know today. By 5500 B.C. there were large lowland villages dependent on irrigation agriculture; and interestingly, there was also evidence of some degree of craft specialization in pottery and metalworks. Paralleling this, settlements in upland areas increased in size, number, and complexity. A wealth of archeological material from one such site, Catal Hüyük of central Anatolia, attests to the sophistication and complexity of life in this early period.[31] Dating from around 6250 B.C. Catal Hüyük was one of the largest settlements of its day (about 13 acres), possibly containing several thousand inhabitants dwelling in well-consructed houses containing partitioned rooms, windows, platforms for sleeping, hearths, and ovens. Many houses also contained burials, polychromatic wall paintings, domestic religious shrines, clay figurines, elaborately painted pottery, and even clay stamp seals. The evidence for basketry, woven goods, and tool manufacturing suggests that some households may also have engaged primarily in craft industry. The community displays signs of both internal social differentiation and a high level of religious and ritual elaboration. There are signs of extensive trade relations with other regions, including contact with the Sinai Peninsula some thousand

[30] Adams, *Heartland of Cities.*

[31] James Mellaart, *Catal Hüyük: A Neolithic Town of Anatolia* (New York: McGraw-Hill, 1967).

kilometers away. The economy that sustained this cultural development was one that combined irrigation farming, animal husbandry, and trade.

The shift to irrigation agriculture in the lowland areas and river valleys, paralleled as it seems to have been by an increase in overall population and by the development of large, dense settlements, was the prelude to the evolution of urban centers and the earliest empires. Thus, by 3000 B.C. the cultural canvas of the Middle East took on the special texture of complexity and local contrast that still distinguishes it today. Also there is some evidence from this early period of a way of life often considered the most distinctive in the area—that of *nomadic pastoralism*.

It is impossible to document fully the rise of this way of life in which humans came to rely on a highly specialized and often risky undertaking. All we know for sure is that while animals were domesticated at about the same time plants were elsewhere, specialized nomadic pastoralism is a later development, probably accompanying the shift to lowland settlement and urbanism.[32] One suggestion is that changes in agricultural practices, especially in increased emphasis on canal irrigation, created the preconditions for specialized herding.[33] Canal irrigation increased productivity, which led to an increase in population and an expansion of settlements. This meant that, in a practical sense, many villages were increasingly established in arid locales, remote from areas of lush grazing. It is also possible that with the intensification of agriculture, the land available for grazing became more and more limited. To get adequate food for their animals, herders would probably have had to travel greater and greater distances. This involved an appreciable investment of time and labor, and could have conflicted with their agricultural pursuits, which with irrigation had become a time-consuming activity. In addition to the work of planting and harvesting, canals had to be dug, cleaned, and repaired, and the allocation of their water monitored. These conflicting demands for time and labor could have encouraged certain households, and ultimately larger groups, to specialize in increasingly intensive agriculture, while others devoted most of their attention to animal husbandry.

From this perspective then, pastoralism arose as a by-product of intensive agriculture. Even though this hypothesis cannot be confirmed, we know that pastoralism in the Middle East is a strategy predicated on agricultural surplus and is closely linked economically with farming communities. Everywhere in the Middle East, for example, the diet of the pastoral household is based on grain or dates as a primary staple.

[32] Flannery, "The Ecology of Early Food Production in Mesopotamia," pp. 1254–55.

[33] Susan H. Lees and Daniel G. Bates, "A Systemic Model for the Origin of Specialized Nomadic Pastoralism in the Near East," *American Antiquity*, 39 (1974), 187–93.

THE RISE OF CIVILIZATION

The Neolithic set the stage for the rapid transformation of the Middle Eastern cultural landscape. In fact, by 3000 B.C. we move from the domain of prehistory to the era of early civilizations, with their written records and monumental architecture.

When we read written accounts from Mesopotamia or Egypt, or when we look at the scenes depicting everyday life later painted on the temple walls of Luxor and other Egyptian sites, the impression we get may be exotic in its details but it is remarkably familiar in overall tenor. What is depicted are different aspects of that cultural complex referred to as *civilization*—a way of life that was qualitatively different from anything else that had preceded it. We see evidence, for example, of the presence of such nonagricultural occupations as artisan, merchant, priest, soldier, and king. We also see the division of society into a number of social classes and the political dominance of a few large centers over an extensive hinterland. Long-distance trade involving luxury and subsistence foods increased in importance.

But of all the features that characterized civilization, the two most significant were the development of large, dense settlements—the basis for *urbanism*—and the emergence of centralized political institutions—the basis for the *state*. Just as the city marks a new form of human settlement and social life in the Middle East, the state marks a transformation in the political order. The state as a form of political organization involves much greater concentration of political power and the presence of a specialized administrative hierarchy.

Although we may never fully understand the genesis of civilization in the Middle East, nor the specific developments that led to the formation of the state, what we do know is that by 3000 B.C. patterns of life in the area had acquired a form whose contours persist even today.[34] This complex can be described in terms of a number of central themes, such as the relationship between the rulers and the ruled, religion and the state, the family and the community, the farmer and the town dweller. What is remarkable is that we have documents from these early periods

[34] For recent discussions of the origin of states in the Middle East, see, for example: Gregory A. Johnson, *Local Exchange and Early State Development in Southwestern Iran*, Museum of Anthropology, University of Michigan, Anthropological Papers No. 51 (Ann Arbor: University of Michigan, 1973); Harvey Weiss, "Periodization, Population and Early State Formation in Khuzistan" in *Mountains and Lowlands*, ed. Louis D. Levine and T. Cuyler Young, Jr., Bibliotheca Mesopotamica 7 (Los Angeles: Undena, 1977), pp. 347–69; Henry T. Wright, "Recent Research on the Origin of the State," *Annual Review of Anthropology*, 9 (1977), 379–97; Henry T. Wright and Gregory A. Johnson, "Population Exchange and Early State Formation in Southwestern Iran," *American Anthropologist*, 77 (1975), 267–89.

Mudifs **on Sumerian cylinder seals, c. 300 B.C.**

A *mudif* of the Beni Isad, Ech-Chibayish.

Guest houses of antiquity and of today in lower Mesopotamia. (Drawing courtesy of Kathleen Borowik)

written on clay that give us a sense of the life at that time. Among these documents are letters—from father to son, governor to king, husband to wife, steward to overseer—all painting a vivid picture of social life. We have chosen the following excerpts because of their intrinsic interest and human touch.

The first excerpt, taken from a Sumerian text of approximately 2500 B.C., recounts a conversation between a father and son. The father begins by asking his son:

> *"Where did you go?"*
>
> *"I did not go anywhere."*
>
> *"If you did not go anywhere, why do you idle about? Go to school, stand before your 'school' father (professor), recite your assignment, open your schoolbag, write your tablet, let your 'big brother' write your new tablet for you. After you have finished your assignment and reported to your monitor, come to me, and do not wander about in the street. Come now, do you know what I said?"*
>
> *"I know, I'll tell it to you."*
>
> *"Come, now, repeat it to me."*

"I'll repeat it to you."

"Come on, tell it to me."

"You told me to go to school, recite my assignment, open my schoolbag, write my tablet, while my 'big brother' is to write my new tablet. After finishing my assignment, I am to proceed to my work and to come to you after I have reported to my monitor."

"Come now, be a man. Don't stand about in the public square or wander about the boulevard. When walking in the street, don't look all around. Be humble and show fear before your monitor. When you show terror, the monitor will like you."[35]

As is clear from this conversation, education, in this case probably for the post of a scribe, was already the responsibility of the school and specialized teachers. It also seems clear that city youth, then as now, need not necessarily have appreciated the opportunities their fathers provided for them. We might also note that the respect and formal deference expected then of a pupil to a teacher is still very much in evidence in the Middle East today.

Like education, medicine too was a specialized craft—one for which the Middle East has long been famous. In Sumer, the physician was known as *a-zu*, or "water thrower", and one text from this time records the following treatment, thought to be for venereal disease:

Having crushed turtle shell and . . . , and having anointed the opening [of the sick organ, perhaps] with oil, you shall rub [with the crushed shell] the man lying prone [?]. . . After rubbing with the crushed shell you shall rub (again) with fine beer; after rubbing with fine beer, you shall wash with water; after washing with water, you shall fill (the sick spot) with crushed fir wood. It is (a prescription) for someone afflicted by a disease in the tun *and the* nu.[36]

The *tun* and *nu* have yet to be precisely identified, and the efficacy of the treatment is unknown. Beer, however, apart from its medicinal properties, was widely consumed and may have been the major impetus for grain production. The fermentation of beer from wheat and barley was a primary way of utilizing these cereals as food. Today, however,

[35] Translated by Michel Civil, cited in Samuel Noah Kramer, *The Sumerians: Their History, Culture and Character* (Chicago and London: University of Chicago Press, 1963), p. 244.

[36] Ibid. p. 99.

the Moslem inhabitants of the Middle East are prohibited by their religion from consuming alcoholic beverages, although in practice many do.

THE CULTURAL HERITAGE OF EARLY HISTORY

The two preceding excerpts refer to life in Sumer, but they could easily have come from Akkad, Babylon, Assyria, or even Egypt—all of which have left their indelible mark on the human and cultural makeup of the Middle East.

The history of dynastic Egypt is usually traced back to 3100 B.C., when Menes, the king of Upper Egypt, conquered Lower Egypt and ruled the united country from his new capital of Thebes. He, like subsequent pharaohs, was believed to be the living embodiment of the falcon-god Horus and was therefore considered divine. A succession of dynasties ruled Egypt until its conquest by the Romans. Presiding over an elaborate priesthood and a bureaucracy comprised of officials, scribes, and clerks, Egyptian pharaohs generally succeeded in preserving the unity of the country despite its many invasions. These millennia of unity no doubt contributed to the Egyptians' strong sense of national identity, which transcends the claims of Islam and pan-Arab nationalism.

According to tradition, Christianity was first introduced into Egypt by St. Mark. It was not until the fifth century, however, that a separate Coptic national church was established as a result of schismatic movement within the Byzantine Church. The Moslem invasion of Egypt that began in A.D. 639 brought many Arab settlers to the country; it also resulted in massive conversions to Islam and the spread of the Arabic language. The result is that the Christian Copts constitute a minority in Egypt today—between 5 and 7 percent of the total population.

Unlike the Nile Valley, Mesopotamia experienced a far more varied cultural and dynastic history. Lacking geographic unity and the easy communication of the Nile, Mesopotamia, or the land between the Tigris and Euphrates rivers, was the scene of a succession of city-states, which included those of Sumer, Akkad, Babylonia, and Assyria. With the exception of the Sumerians, the inhabitants of these states spoke related languages classified as Semitic, a grouping that today includes modern Arabic and Hebrew.

Other Semitic-speaking populations, such as the Phoenicians and the Canaanites, settled in the eastern Mediterranean areas astride the cultural frontiers of the Egyptian and Mesopotamian civilizations. Seafarers and traders, the Phoenicians established their cities throughout the Mediterranean and in time came to develop a system of writing based on a phonetic alphabet, which replaced earlier hieroglyphics and cuneiform.

The ancient Hebrews trace their origins to Mesopotamia, from which they migrated to the eastern Mediterranean and later to Egypt. Following their exodus from Egyptian bondage in the thirteenth century B.C., the Hebrews succeeded in conquering parts of the lands of Canaan where they settled. Under King David (1000 to 960 B.C.) the various Hebrew tribes united, forming a state with its capital at the newly captured Jerusalem. Following the death of Solomon, however, the kingdom divided into two nations—Israel in the north and Judea in the south. Israel was destroyed by the Assyrians in 721 B.C., but the smaller kingdom of Judea survived until 586 B.C., when it fell to the Babylonians; they exiled its leaders to their capital of Babylon in what is today southern Iraq. In 539 B.C. the Persian Emperor Cyrus conquered Babylon; a year later he allowed the Jews to return to Jerusalem and to rebuild their temple. The Jews remained in Palestine, maintaining a semblance of independence under a succession of Persian, Greek, and Roman rulers until 63 B.C., when the Romans finally destroyed the temple.

Another Semitic-speaking population is the Arabic one. The term *Arab* first appears in an Assyrian cuneiform tablet dating from 853 B.C. and seems to have referred to nomadic pastoralists.[37] With time, the term acquired its most common usage, designating the speaker of one of a group of closely related dialects of the Arabic language, which is itself divided into two major groupings, northern and southern Arabic. Dialects of southern Arabic survive today in scattered islands in the Indian Ocean, but few speech communities remain anywhere on the Arabian Peninsula.

From ancient days, Arab populations were active in land and sea commerce with trading communities found as far as China, India, and East Africa. Distributed over a broad area extending from the steppes of Syria to the highlands of Yemen, Arabic speakers exhibited great diversity in political and social organization. For example, one group, the Nabatean, established a prosperous trade center, Petra, whose elaborate buildings were carved into the pink sandstone of the cliffs in today's Jordan. To the east in the Syrian steppe, Palmyra was another important Arab city-state whose impressive architecture stands abandoned in the desert. Its last ruler, Queen Zenobia, managed to extend her realm until meeting final defeat at the hands of the Romans in A.D. 273.

In the period that preceded the Islamic conquest, a number of small Arab states were established along the buffer zones that separated the

[37] Bernard Lewis, *The Arabs in History*, 2nd ed. (New York: Harper and Row, 1967), pp. 10–17.

two dominant empires of the time—the Byzantine and the Persian Sassanians. As clients of one or the other, Arab kingdoms such as the Christian Ghassanids of Syria and the Lakhmids of southern Iraq survived until absorbed by the Persian Empire.

The Persians, who speak an Indo-European language, are heirs to a great civilization and a powerful empire that at one time stretched from Asia Minor to India. The Sassanian rulers of Persia, the last of the non-Moslem dynasties, had adopted Zoroastrianism as the national religion. It remained so for a period of about 1000 years, until the Persian defeat by the Moslem Arabs. Even so, the conversion of Zoroastrians to Islam proceeded slowly in many places, and even today a small Zoroastrian community can be found in Iran. Zoroastrianism, which emphasizes the coexistence of good and evil and humanity's responsibility to uphold the good, is believed to have strongly influenced the development of early Judaism and, as a consequence, Christianity and Islam in turn.

THE ISLAMIC CONQUEST AND AFTER

The Arab-Islamic conquests that began in the seventh century ushered in a new phase in the development of the cultural history of the Middle East. These conquests initiated a process of cultural synthesis that culminated in the formation of an Arab-Islamic civilization. The synthesis is based on the spread of the Arabic language and the religion of Islam. The conquest itself proceeded very rapidly, sweeping through the Middle East in the course of very few years. What is particularly striking is the rapidity with which Islam as belief and polity established itself in the heartland of ancient civilizations. One way this can be understood is to realize that in the period preceding Muhammad's birth, both the Byzantines and the Persians were greatly weakened and their claims on the local populace had eroded. Islam then arose and spread in what may be considered a political vacuum, at least inasmuch as one thinks of strong centralized state rule.

Of course, the advent of Islam, which arose in the Arabian Peninsula, did not result in immediate conversions of the masses of people who came in contact with the Moslem armies. The Arab conquerors themselves often discouraged the conversion of non-Arabs, seeking to preserve for themselves the privileged status of a conquering elite. However, Islam, as a universal religion, acknowledges no ethnic boundaries—a fact that in time encouraged the conversion of Aramaic-, Persian-, and Greek-speaking populations. With the exception of the Persians, the converted population fairly rapidly adopted Arabic, the language of government as well as religion. Even in Persia, the educated classes wrote in Arabic, and Arabic script came to be employed for writing Persian itself. At the same time, some populations, while adopting the Arabic language, nonetheless retained their separate religious identities; Christian and Jewish communities persisted within the Islamic order.

One important population which, while embracing Islam, retained its language is the Turks, whose language belongs to the Altaic group (which also includes Mongolian). Chinese chronicles refer to groups of presumably Turkish nomadic pastoralists in central Asia as early as 1300 B.C. But the appearance of Turkish speakers in southwestern Asia is relatively recent, dating from about the ninth century A.D. Initially, Turkish mercenaries or warrior-slaves were recruited to the service of imperial houses of the Fatimids of Egypt, and Abbassids of Baghdad, and other dynasties. With time, these caste-like military contingents gained considerable power in their own right, and their ranks swelled as Turkish-speaking populations migrated westward from central Asia in increasing numbers. It is usually thought that this movement of people was set in motion by the Mongol conquests. Although the processes by which the Turks established themselves as a major political and cultural presence are still unclear, historians often identify two main waves of conquest and political consolidation. Each period of consolidation was made possible because diverse Turkish-speaking groups or tribes had already migrated into the areas in question.

The first period of Turkish political rule over a major state or empire occurred when the Seljuk Turks of Persia gained ascendancy in the tenth century A.D. over most of the former Abbassid domains. By A.D. 1071, Seljuk rule and a Seljuk capital were soon established in the city of Konya, formerly Aykonium. Following A.D. 1071, when the Byzantines were decisively defeated in the east, Turkish settlers, both as nomadic tribes and warriors, arrived in large numbers in what is today Turkey. Ultimately, Seljuk rule was broken by the Mongols and power was fragmented among numerous small Turkish emirates or ministates, some no more than tribal confederations.

In the midthirteenth century, one of these emirates, the House of Osman, succeeded in establishing hegemony over its rivals and expanded its territory at the expense of the remaining Byzantine lands. This marked the onset of the second period of Turkish political and cultural expansion, that of the Ottoman Turks. By the fourteenth century, the Ottomans were masters of the whole of Asia Minor with the exception of Constantinople, and they had crossed the Bosphorus to acquire substantial territories in Europe. Where the early Arab armies failed in the seventh century, the Turks succeeded in capturing Constantinople, renamed Istanbul in 1453. This removed the last remnants of the Byzantine political presence from the area.

At its apogee, the Ottoman Empire stretched from the great plains of Hungary in the west through the Balkans and around the Black Sea, continuing in a vast arc to encompass virtually the entire eastern and southern Mediterranean. With the exception of Iran, the countries we are concerned with in this book are recent successor states to the Ottoman Empire, whose legacy is still visible. For over three centuries, the Ottoman Empire, one of the world's major imperial systems, was also the dominant power in Europe and remained a political force in the Middle East until its dissolution at the end of World War I.

In many ways the Turkish experience exemplifies the ideological appeal of Islam and its role in the political and cultural domains. Early Turkish warriors, adopting Islam, consecrated themselves to the task of extending its frontiers. Achieving power in their own right, Turkish dynasties utilized and adapted Islamic ideology and institutions to legitimize their rule and hold together a vast, multiethnic empire.

An almost parallel case is to be found with the long-time rivals of the Ottomans to the east, the Safavids of Persia. In 1501 a Safavid Shah proclaimed Shi'ism, a distinctive form of Islam, as the state religion, and consolidated power over a smaller but also an ethnically diverse empire. Despite long rivalry with the Ottomans, Persians successfully resisted Ottoman encroachment to the east of the Zagros.

We could mention a number of other populations and dynasties, both long established and relatively recent, that contributed to the mosaic that is the Middle East today. But the point we wish to emphasize here is that Islam was adopted and adapted by diverse peoples over a long period of time. As a consequence of these processes, Islamic values, ethos, and institutions contributed to a strongly shared tenor of life and common cultural experience for the peoples of the Middle East. However diverse the local cultures of the Middle East, Islam remains a major source of shared identity. The two following chapters both introduce Islam and explore the ways in which it is interwoven into the social and cultural fabric of the region.

CHAPTER TWO: ISLAM: THE PROPHET AND THE RELIGION

In the name of God, the Merciful, the Compassionate,
Praise be to God, Lord of the Universe,
The Merciful, the Compassionate,
Ruler on the Day of Judgement.
Thee alone we worship, Thee alone we ask for aid.
Guide us in the straight path,
The path of those whom thou has favored,
Not of those against whom thou art wrathful,
Nor of those who go astray.

This prayer, the *fatiha* or opening chapter of the Quran, the Holy Book of Islam, is one of the world's most often recited sacred verses. Of the 600 million people who profess Islam, a great number direct this praise to God four times before each of their five daily prayers, as well as before embarking on any important task or journey. Uttering these words, a traveler setting out from Fez in Morocco on a journey eastward to Afghanistan will pass through countries that differ in climate, language, and customs, but everywhere he or she will be identified as a member of the universal community of the Faithful, or the *Umma*. In each of the countries on the way, the traveler will hear the public call to prayer chanted by the *muezzin* from the minarets of mosques or, more

A mosque from Edirne, Turkey, European gateway to the Middle East. (Photo courtesy of Ulku Bates)

likely today, broadcast by tape recorders through loudspeakers. The call that beckons the believers to prayer proclaims the central article of faith for all Moslem peoples: "I profess that there is no god but God and that Muhammad is his Messenger." This simple statement, which constitutes the declaration of faith, is regarded as the most fundamental part of the Islamic creed. Its threefold public recitation in Arabic is, for all intents and purposes, sufficient to make one a convert to Islam. Having spoken these words in seriousness, one becomes a member of that vast community, the *Umma,* subject to its laws and recipient of its support.

THE SCOPE OF ISLAM

Despite evident diversity in both expressed belief and observed practice, those individuals who profess Islam thereby proclaim their membership in a community that transcends ethnic and national boundaries. The shared sense of one Islam, eternal and immutable, is itself a distinguishing and fundamental characteristic of the faith, for it irrevocably sets apart those who have accepted God's final prophecy from those who have not. Islam is a universalist religion like Christianity and Buddhism to which every person can belong. It is, in the words of Maxime Rodinson, one of the great "ideological movements" in world history. It has created a new community and endowed its members with a distinct identity; at the same time, it is an ideology projecting an ideal society.

When Islam is viewed sociologically, it is evident that it draws on many sources of belief and practice. Indeed, it is a futile venture to attempt a universal definition of the beliefs and practices of any living religion. Islam, like all religions that claim universal validity, is best viewed as an ongoing, ever-changing, living tradition. One aspect of this complex tradition in Islam is that of a set of beliefs and history recorded in scriptures that are passed on and reproduced from generation to generation. This fact cannot be overstressed if one is to understand the role of Islam in Middle East societies. This written repository of belief and history is by and large the domain of religious scholars and the learned, who as such come to exercise considerable power and authority as well as provide a major source of cultural continuity. Islamic scripture and its scholarly interpretations might be thought of as constituting the formal expression of the Islamic tradition.[1]

In analyzing complex literate societies, it is often useful to distinguish the systems of belief and practice of the learned or the elite from the understandings of the common people. The perceptions of Islam by the learned and religious specialists will vary by region and sect, perhaps even making it impossible to establish a single shared dogma; still, the Islam (or Islams, as some have said) of the learned displays less variation than do the beliefs and practices of the common people. This analytic distinction is often described as one between the Great or Universalistic Tradition and the Little or Particularistic Tradition. Moslems themselves continue to debate "orthodoxy" and what constitutes "true" belief and practice, but the analytic fact remains that variation in practice and interpretation is inevitable in a living religion. How people understand, interpret, and act upon Islamic principles defines what Islam is at any given time for a particular community.[2]

[1] Readers interested in learning about the formal expression of the Islamic Tradition might consult various entries in the H. A. R. Gibb and J. H. Kramer, eds., *Shorter Encyclopedia of Islam* (Ithaca: Cornell University Press, 1953). For a good overview of Islam and Islamic institutions, see W. C. Smith, *Islam in Modern History* (New York: New American Library, 1957). A well-written general introduction to the subject is Fazlur Rahman, *Islam*, 2nd edition (Chicago: University of Chicago Press, 1979).

[2] A review article of anthropological approaches to the study of Islam is Abdul Hamid el-Zein's, "Beyond Ideology and Theology: The Search for the Anthropology of Islam," *Annual Review of Anthropology*, 6 (1977), 227–54. For examples of local-level studies of

Clearly, each of these two aspects or traditions of Islam informs the other. Formal or scriptural Islam can be seen at any one time as the source of an ideal code and as a set of notions against which the reality of human behavior can be measured. It is, in fact, as we shall see later, a primary source for the law. At the same time, this ideal code or formal system of belief itself reflects an ever-changing experience. This chapter attempts to describe some of the basic tenents and beliefs that lie at the heart of Islam. In the subsequent chapter, we shall explore the diversity in belief and practice which is also Islam and which justifiably can be termed Islamic culture. One point to remember is that the origin and spiritual roots of Islam are in the Middle East; it, more than any other factor, defines the area culturally.[3]

THE RISE OF ISLAM

One of the three great monotheistic religions of the world, Islam arose in the full light of history. Developing six centuries after Christianity, it rapidly achieved astounding success. At the time of his death in A.D. 632, its founder, the Prophet Muhammad, was the undisputed ruler of most of Arabia. "Moreover, within ten years of his death the state which he created was able to meet in battle and defeat the armies of the great empires of the Middle East, the Byzantine and the Persian, and within a short time to overrun the latter completely. A hundred years after his death, the empire of his successors extended from France to India."[4] Continuing to gain adherents, Islam is even today the most rapidly expanding religion in the world, especially on the African continent.

In order to understand and appreciate Islam as a religious, political, and social force, we turn our attention briefly to sixth-century Arabia, the birthplace of Muhammad. Like other ideological systems, religions evolve in specific economic and political contexts. They not only reflect the social tensions of the moment, but also themselves shape

Islamic tradition, see Dale Eickelman, *Moroccan Islam: Tradition and Society in a Pilgrim Center*, Middle East Series, 1 (Austin: University of Texas Press, 1976); Clifford Geertz, *Islam Observed* (Chicago: University of Chicago Press, 1968); and John Mason, *Island of the Blest: Islam in a Libyan Oasis Community* (Athens, Ohio: Ohio University Press, 1977).

[3] See, for example, G. E. Von Grunebaum, *Islam: Essays in the Nature and Growth of a Cultural Tradition*, Washington, D.C.: Memoirs, The American Anthropological Association, 1955).

[4] W. Montgomery Watt, *Muhammad: Prophet and Statesman* (London: Oxford University Press, 1961), p. 5. A good, if somewhat idiosyncratic, general treatment of the origin, spread, and development of Islam is to be found in Marshall Hodgson, *The Venture of Islam: Conscience and History in a World Civilization* (Chicago and London: University of Chicago Press, 1974).

ongoing processes of change. Islam has its origins in the pervasive social and economic transformations that were taking place in Arabian society in the sixth century. At the same time, as a social and political force in its own right, it contributed to the transformation of Arabian society and eventually to areas far beyond it.

The Arabian Peninsula connects the lands of the Levant with the Indian Ocean, and from antiquity has served as the crossroad between the great empires of the Mediterranean and the Far East. Although little is known about the people of central and northern Arabia, it appears that many were nomadic pastoralists and that they were organized in tribes and confederations that sometimes united nomad and oasis dweller together. Local resources alone were probably not sufficient to sustain the development of larger polities like the small-scale states and kingdoms which had developed earlier in southern Arabia.[5]

The Romans, who dominated the Levant in an earlier period, rarely bothered to establish their direct rule over Arabia, but were generally content to exert their influence indirectly through control of the many small client states or chiefdoms that arose, prospered, and declined with regularity along the desert frontier of northern Arabia. As we noted earlier, two of the more famous of these kingdoms were Petra and Palmyra.

The Romans were superseded by the Byzantines, but the general political pattern in the area remained much the same until the fourth century. By this time a number of Arab tribes had converted to Monophysitic Christianity, while the majority retained their earlier beliefs. In the next two centuries there appears to have been wide-scale economic deterioration and general political upheaval in the Arabian Peninsula. The exact causes still remain obscure, but around this time the monarchies of southern Arabia collapsed and their agrarian economies fell into ruin. Some of these kingdoms fell to the Persian invaders, and others succumbed to the Abyssinians. Southern tribes subsequently embarked on a series of migrations to the north that brought them into conflict with each other and with the northern tribes. The once-prosperous trans-Arabian trade languished, and the routes fell into disuse.

By the sixth century A.D., a long period of local economic deterioration and intertribal warfare had worked itself out and a relatively stable pattern emerged. Many of the Arabian nomads had settled down on oases, some founding new towns in the process. Other oases were inhabited by Arabic-speaking Jewish populations, while Christians were not uncommon among some of the nomadic tribes. Christian monasteries were scattered throughout the northern part of the Peninsula, where some, such as St. Catherine's in the Sinai, remain today.

[5] Bernard Lewis, *The Arabs in History* (New York: Harper and Row, 1966), p. 24.

Trade was again becoming important, and the local populations played important roles as caravaneers, middlemen, and merchants. Towns along the major caravan routes grew wealthy. Among the most prosperous of the new towns was Mecca, which had been founded around A.D. 100 by the Quraish, a northern Arabian tribe, around the well of Zamzam. By the fifth century Mecca had become the major trade town along the western coast of Arabia (known as the Hijaz), and its merchants maintained commercial relations with both the Byzantines and the Persians. The town itself seems to have been ruled by an oligarchy made up of the leading merchants, most of whom belonged to the Quraish.

Besides long-distance trade, another source of revenue for the Meccans was the local shrine, a large, pantribal sanctuary that housed the images of the many gods and goddesses worshipped by the Arabs before Islam. It is said to have even contained some Christian and Jewish relics as well. The most sacred object in the sanctuary was the *Ka'ba,* a black stone that was part of a meteorite, considered holy by the different tribes who came to Mecca to worship at the shrine and to attend the busy market nearby. The fame and success of this market, known as *suq 'ukaz,* was in no small measure due to the presence of the sanctuary. The sanctuary of Mecca, one of several in the Peninsula, was considered a sacred place, or *haram*—a consecrated area where no blood could be shed and where oaths could be taken. The sanctity of the shrine no doubt extended to the market area to ensure trust in business transactions and to guarantee a temporary truce among the chronically feuding tribes. The Quraish elders, in their capacity as the elite of Mecca, controlled and derived revenues from both the sanctuary and the market.

MUHAMMAD, THE MAN AND HIS PROPHECY

Muhammad, a member of the Quraish tribe, was born around A.D. 570. His family was fairly impoverished, and his mother, who came from Medina, a town to the north of Mecca, died when he was 6. He was brought up first by his grandfather and then by his paternal uncle Abu Taleb, a wealthy merchant and a respected member of the Quraish oligarchy. Abu Taleb discerned intelligence and initiative in the boy and employed him to accompany his caravans as they traded in the north. It was probably on these journeys that Muhammad came to meet the elders or scholars from the several Arabic-speaking Jewish and Christian communities that were found in Syria and Arabia, and through them learned something about their beliefs.

There are few authentic stories about Muhammad's early years before his apostleship. One tells of the encounter between Muhammad and a man named Zayd, who apparently was banished from Mecca for preaching some kind of monotheistic belief. The story is related by Muhammad's first biographer, Ibn Ishaq, who wrote:

> *I was told that the Apostle of Allah said as he was talking: "I had come from Al-Ta'if... when we passed Zayd son of 'Amr who was in the highland of Mecca. The Quraish had made a public example of him for abandoning their religion, so that he went out from their midst. I sat down with him. I had a bag containing meat which we had sacrificed to our idols ... and I offered it to Zayd—I was but a lad at the time—and I said 'Eat some of this food, my uncle.' He replied, 'Surely it is part of those sacrifices of theirs which they offer to their idols?' When I said that it was, he said, 'Nephew of mine, if you were to ask the daughters of 'Abd al'Muttalib they would tell you that I never eat of these sacrifices, and I have no desire to do so.' Then he upbraided me for idolatry and spoke disparagingly of those who worship idols and sacrifice to them, and said, "They are worthless; they can neither harm nor profit anyone," or words to that effect." The Apostle added, "After that I never knowingly stroked one of their idols nor did I sacrifice to them until God honored me with his apostleship."*[6]

By the time he was 20, Muhammad had acquired a reputation for wisdom and trustworthiness. These qualities apparently brought him to the attention of a wealthy widow, Khadija, who hired him to manage her caravans and supervise her business. She eventually proposed marriage, and at age 25 Muhammad married Khadija, who was 15 years his senior. The marriage seems to have been a happy one; Khadija bore him a number of children, including a favorite daughter, Fatima, who later married Muhammad's first cousin 'Ali. The descendants of this latter marriage, called *Sayyids*, are greatly revered by Moslems all over the world today.

Although little is known about the period immediately preceding the apostleship of Muhammad, scholars have nevertheless attempted to understand the development of Muhammad's prophecy and career in terms of the socioeconomic transformations of his day. Of the Western scholars, we single out W. Montgomery Watt and Maxime Rodinson, who are concerned with showing the relationship between Muham-

[6] Alfred Guillaume, *Islam*, 2nd. ed. (Penguin Books, 1956), p. 26.

mad's mission, the success he had in acquiring a following, and the prevailing social and economic conditions in western Arabia. In their view, once launched as a distinct religion, the transformation of Islam into a political movement was historically inevitable. Although it developed in the context of a tribal society rent by factions, Islam, as a universalistic movement, managed to transcend these cleavages and restructure the society along new lines.[7]

In Rodinson's interpretation, Muhammad's early dissatisfaction with the pagan religious practices of his fellow Arabians and with the wide differences in wealth within Meccan society are closely related. At this time Mecca included not only the wealthy merchant families of the Quraish, but also their dependent clients, slaves, and the newly settled nomads who made up the majority of the inhabitants. The disintegration of tribal cohesion and the growth of social differentiation must have become particularly accelerated. As individuals became wealthy from trade, the traditional tribal norms of mutual aid and protection increasingly fell into disuse; the poor and the powerless began to be abandoned by the clan and were left out of its protective network. The old values that operated in an egalitarian, nomadic group were being superseded by values that stressed individualism, material display, and competition. In fact, rich merchants in Mecca joined together to form commercial associations whose object was to monopolize trade and keep away rivals. Loosely organized along clan and tribal subdivisions, these mercantile associations also functioned as political factions as they competed for the right to manage and control the pilgrimages, fairs, and trading activities of the city.

Impressed by growing social differentiation within Mecca and sensitive to the plight of its needy and neglected groups, Muhammad took to retreating from the city to the nearby mountains to meditate. It was not unusual in Arabia for individuals to retreat to mountain caves and deserts in order to meditate in solitude and to ponder their spiritual and psychic problems, and Muhammad's action seems to have aroused no curiosity or concern, at least not initially.

It was in one of the caves on the nearby mountain of Hira that Muhammad first underwent a profound religious experience in which he believed he was called to become God's messenger, charged with revealing the Truth to humanity. The year was A.D. 610, when Muhammad was already 40 years old. That experience marked the beginning of his Prophetic career.

Tradition has preserved the details of this first experience. Alone in the cave, Muhammad began to hear voices and see visions. Later he saw

[7] W. Montgomery Watt, *Muhammad*, and Maxime Rodinson *Muhammad* (New York: Pantheon Press, 1971).

Interior view of the Mosque of Nihrihmah Sultan, Istanbul. (Photo courtesy of Ulku Bates)

an apparition which he identified as the Archangel Gabriel. The Heavenly Messenger commanded the frightened man to speak, but Muhammad refused. Gabriel repeated the command three times, and at the third command, Muhammad spontaneously recited the following verses:

> Recite (*iqra'*): In the name of Thy Lord who Created,
> Who created man from a blood-clot,
> Recite: And thy Lord is the most generous, who taught by the pen
> Taught man that he knew not.

Thus began Islam and the Quran, or Holy Book, which contains Muhammad's revelations, which spanned a period of approximately 22 years. The word *Quran* is derived from the first word of the first revelation, *iqra'*, which is from the Arabic root *qra'*, meaning *to read* or *to recite.* The early revelations were received by Muhammad in rhymed prose, a form of recitation widely used in Arabia by poets and soothsayers.

This religious or mystical experience was followed by others, and Muhammad slowly came to accept his role as God's Apostle, the one chosen to receive and preach God's word. The first to believe in him were members of his immediate family, notably his wife and his first cousin, 'Ali. For three years Muhammad limited his preaching and conversions to a small group of intimates who used to meet in secret. Then he decided to preach in public, and the group met daily to hold prayers, an activity that quickly brought them to the attention of the Meccans. At first the Meccans simply mocked Muhammad and his followers, but when it became apparent that more and more people were joining Muhammad's circle, the Meccan elite took steps to put an end to a movement that threatened their position as guardians of the holy sanctuaries and rulers of the city. By this time Muhammad's followers included a number of young wealthy men, but the majority were from among the weak, the poor, and the powerless of the town.

The sanctions undertaken by the Quraish included the harassment of the Moslems and the boycott of Muhammad's clan of Hashem. This meant that the other Quraish clans refused to intermarry with the Hashem or to have any business dealings with them.

> *The resistance of the Meccans appears to have been due not so much to their conservatism or even to religious disbelief (though they ridiculed Muhammad's doctrine of resurrection) as to political and economic causes. They were afraid of the effects that his preaching might have on their economic prosperity, and especially that his pure monotheism might injure the economic assets of their sanctuaries. In addition, they realized . . . that their acceptance of his teaching would introduce a new and formidable kind of political authority into their oligarchic community.*[8]

[8] H. A. R. Gibb, *Mohammedanism: An Historical Survey* (New York: The American Library, 1958), p. 30.

In A.D. 619 Abu Taleb, Muhammad's uncle and protector, died and was succeeded to the clan's leadership by another relative, who was not well disposed towards Muhammad. In fact, he withdrew the clan's protection from Muhammad and his followers, which made it dangerous for the Moslems to stay in Mecca. Three years later, in A.D. 622, Muhammad and some seventy of his followers migrated to Yathrib (later Medina), an oasis town some 200 miles to the northeast of Mecca.

UMMA: THE ORIGINS OF THE ISLAMIC POLITY

The migration, or *Hijra,* marks a new phase in the evolution of the Islamic community. In Medina the religious movement was soon embodied in a political form, that of the *Umma,* or community. In fact, the migration was considered so important by the Moslems that the first day of the year in which it took place, July 16, A.D. 622, marks the start of the Islamic calendar, in which a year is based on twelve lunar months.

Unlike Mecca, which depended on commerce for its livelihood, Medina was an agricultural town where dates and cereals were grown. It was inhabited by a number of pagan Arab as well as Jewish tribes who lived in scattered settlements maintaining an uneasy accommodation among themselves. The Jews seemed to have been dominant earlier, but their power had slipped away, and at the time of the migration Medina was experiencing a difficult period of chronic feuding between the different groups. In fact, it was mainly in an effort to put an end to this anarchic and unstable state of affairs that a group of Medina notables had invited Muhammad to their town. They wanted him to act as an arbitor and peacemaker, and in return promised him freedom to preach and asylum for his followers. They were apparently not greatly concerned with his prophecy and religious mission.

In Medina the Islamic movement assumed a new shape—that of a community organized on political lines under the leadership of a single chief. Whereas in Mecca Muhammad and his followers had given rise to a new religious sect, in Medina they forged a state. From this point on, to be a Moslem meant at once to adhere to a faith or religion *and* to be a member of a political community. The dual nature of Islam was thus established early and remains throughout.[9]

Muhammad's first act upon arriving was to regulate Medina's political life. He did this by drawing up an agreement in which the emigrants who came with him and the eight groups already in the town who accepted his teachings were defined as Moslems. These groups

[9] This duality is expressed in the saying, "Islam is at once a religion and a state," *Al-Islam din Wadawla.*

were all conceived as being coequal, their rights and duties were listed, and they were pledged to mutual defense. The pact outlawed bloodshed among Moslems and specified the status of the neighboring Jewish tribes in the area, who as non-Moslems were excluded from the *Umma*. What emerges from this remarkable document is the image of a new confederation of unrelated groups, all primarily united in their common allegiance to Islam with Muhammad acknowledged as their Prophet and leader. This new polity, based on ties of religion rather than those of kinship, was a new development in its Arabian context and formed the nucleus of what later became the Moslem state.[10]

Once established in Medina, Muhammad turned once more to winning Mecca over to his cause. Mecca was the undisputed trading and political center of western Arabia, and its capitulation would greatly enhance the status of Islam as well as guarantee its survival. It took 7 years of skillful combination of military and economic pressures before the Prophet forced Mecca to capitulate. The struggle began as the emigrants to Medina started to finance their own trade caravans, which immediately brought them into conflict with the Quraish. In A.D. 624, over 300 Moslems led by Muhammad ambushed a large Meccan caravan coming back from Syria. This Battle of Badr marked the first military operation by the Moslems. It signaled the transformation of the community into a potent military force, and the political status of the religious community was now made explicit.

The Battle of Badr was followed by a number of skirmishes and engagements with the forces of the Quraish. The consistent success of the Moslems enhanced their reputation among the tribes of Arabia, whose delegations converged on Medina to offer their allegiance to Muhammad and to share in the growing wealth and influence of the Moslem community. By A.D. 630, Muhammad had become the de facto master of most of Arabia, and he was able to enter Mecca with 10,000 of his men. The Prophet proceeded to destroy all the idols at the sanctuary save for the black stone, the *Ka'ba*, which he incorporated into the pilgrimage ritual that makes Mecca the most important of the holy cities of Islam.

Muhammad devoted the remaining 2 years of his life to consolidating the community. This period was one of intensive political activity, as he tried to extend his influence far abroad in Arabia while mediating disputes and rivalries among his followers. He must have succeeded, however, in imparting to his followers his own religious fervor and moral commitment, because following his death, his closest companions quickly became the center of a committed movement that

[10] See W. Montgomery Watt, *Islam and the Integration of Society* (Evanston: Northwestern University Press, 1961).

carried his message beyond the frontiers of Arabia and promoted Islam to one of the world's great religions.

THE REALM OF ISLAM AFTER MUHAMMAD

Muhammad died in A.D. 632 without naming a successor. His closest associates met in council and decided, following tribal custom, to choose one among them to be his deputy, or *khalifa* (caliph). The majority chose Abu Bakr, an old and trusted companion of the Prophet and one of his first converts. But a minority, supporting 'Ali, Muhammad's cousin and son-in-law, insisted that the leadership of the community must remain within Muhammad's family. This group became known as *shi'at' Ali*, or the partisans of 'Ali. The episode heralded the beginning of disunity and factionalism that came to plague Islam and only a few years later was to precipitate its first civil war.

The schism also reflects a fundamental tension between two competing principles of political legitimacy in Islam. On one hand, legitimacy is believed to reside in the will of the community; on the other, it is seen as inherent in rights of descent. In time the minority group—or as they came to be known the *Shi'a*—evolved into the major schismatic division within the Islamic community, a subject that is discussed in the next chapter.

Abu Bakr's first challenge as caliph was to deal with the wave of apostasy that followed the death of the Prophet. Some of the Arab tribes felt that their allegiance to Islam ended with Muhammad's death; they thereupon reneged on their pledges and ceased to observe the ritual and to pay the tax. Moreover, there is evidence to suggest that following the Prophet's death, a number of false prophets appeared in Arabia. The best known of these false prophets was Musaylima, who with an army of about 40,000 men succeeded in defeating the Moslem armies before being himself finally defeated.

Abu Bakr was succeeded by the second caliph, Omar, who presided over the successful expeditions that took the Moslem armies beyond the Arabian frontiers and into battle with Persian and Byzantine armies. Omar (A.D. 634–644), an extremely able administrator, is generally credited with the formation of the system of government that became a model for later Islamic dynasties.

Briefly, soon after the conquest of Byzantine and Persian provinces, the Arab military commanders took over existing governmental institutions, which were kept relatively intact. New cities were founded at some distance from existing population centers to serve as garrisons. This was done in order to consolidate Arab influence in the newly conquered lands, a problem because the Moslem rulers and armies initially were a small minority. The new Arab state appropriated Byzantine and Persian crown lands and the property of important enemy leaders, while explicitly recognizing the property and personal rights of most non-Moslem subjects. In so doing they quickly obtained the acquiescence, if not the active support, of the populace for the new rule. These non-Moslems were, however, required to pay special taxes. This payment evolved into a system of differential taxation for Moslems and non-Moslems which persisted until recently in some countries. Thus, although the Moslem state recognized the rights of non-Moslems (*dhimmis*), the system of taxation greatly encouraged conversion, which proceeded rapidly in most of the conquered areas.

Omar was assassinated in A.D. 644, and a council he appointed chose an unlikely successor, Othman, a member of the Quraish ruling clan. Othman, a man of pious reputation, is recognized today by Moslem historians for reestablishing the influence of the Meccan aristocracy. He regularly placed his relatives in positions of power and ignored the resentment that resulted within the Moslem community at large. Othman was murdered in A.D. 656 and was immediately succeeded by 'Ali, the Prophet's cousin and son-in-law. 'Ali's reign as the fourth caliph marks the end of the caliphate as the expression of consensual leadership of the *Umma;* he is the last caliph recognized by most Moslems as justly elected. His brief rule (A.D. 656–659) was marked by intercommunal dissension and tribal and civil war. One important consequence of this prolonged conflict and 'Ali's ultimate defeat was that the caliphate became dynastic. With this, political power moved out of Arabia.

Mu'awiya, a nephew of Othman and a long-time political opponent of 'Ali, became the first Moslem ruler to found a dynasty, that of the Ummayads. By A.D. 661, after the military defeat of 'Ali, the center of power shifted from Medina to Damascus, which became the new capital of the Ummayad Empire, the first successor state to the early caliphate. With this shift, Islam ceased to be a purely Arab phenomenon limited to the Peninsula. Islam established itself in Damascus at the heart of the Mediterranean world, a successor to the classic empires of Rome and Byzantium; the religion founded by Muhammad had now become the guiding principle and the raison d'étre of a vast, ethnically heterogeneous and urban-dominated empire.

We have dwelt on Muhammad's biography and the historical events following his death because these particulars form the basis for sectarian divisions in Islam, a concern we take up shortly. These events of early Islam, differently interpreted, are continually evoked, even today, to explain and legitimize behavior. For example, the Shi'ite leader

Ayatollah Khomeini of Iran referred to his opponent, the late Shah of Iran, as the "Yazid" of his day, a most powerful Shi'a idiom for expressing tyranny and deceit. The caliph Yazid was the Ummayad ruler charged by the Shi'a as having ordered the murder of Hussain, son of 'Ali and perhaps the most revered martyr of the Shi'a sect. Pageants and passion plays, as well as many of the basic rituals of Islam including pilgrimages, all reproduce these early historical events, giving them great symbolic importance to the believer. They are thus part of the living tradition of Islam.

Against the backdrop of Muhammad's life history and the political developments thereafter, we can now turn to some of the major beliefs and rituals that distinguish Islam.

ISLAM AS FAITH

The word *Islam* means *submission*—that is, the submission of the self to the will of God; it was adopted by Muhammad himself to refer to the distinctive faith he preached, and it appears repeatedly in the *Quran*. A believer in that faith is a Moslem (or Muslim). Needless to say, the ideological system that was initially laid down by Muhammad and that was later interpreted and elaborated by Moslem theologians is too vast and complex to treat in this summary presentation. What follows, therefore, is simply a sketch of the basic principles of Islamic religion, particularly those that distinguish it from the other two monotheistic religions of the area, Christianity and Judaism. At the same time, it must be remembered that Islam draws upon the common Semitic tradition that had earlier produced and nurtured these two predecessors.

Islam has as its central tenet the Oneness of God:

> Say . . . He is God, One,
> God, the Everlasing Refuge,
> Who has not begotten, and has not been begotten,
> And equal to Him is not anyone.[11]

Over and over the Quran preaches strict monotheism; in fact, the worst sin in Islam is to associate other dieties or partners with God. God is conceived as being eternal, omnipresent, and inscrutable; however, this omnipotence is believed to be tempered with justice and compassion. The two basic and most frequent attributes of God in the Quran are the Merciful and the Compassionate.

[11] Sura 112. *The Koran Interpreted*, Vol. 2, trans. A. J. Arberry (London: George Allen and Unwin, 1955), p. 353.

While espousing a strict monotheism, Islamic scriptures also acknowledge the presence of angels; these are pure, sexless beings who dwell in Heaven and who sometimes act as God's messengers. The devil himself is believed to have been an angel, but he was banished from heaven for refusing to obey God's commands. Ranking below the angels and separating them from humanity are a group of male and female spirits, *jinn*, who were created by God from "smokeless fire." These inhabitants of deserts and dark lonely places are mischievous creatures who go around causing trouble; various charms are therefore employed to ward them off.

Moslems believe that God makes his will known to humans through the agency of the prophets who have revealed his commands. He gave the Jews the Torah, the Christians the Gospels, and the Quran to Muhammad. The first of the prophets was Abraham and the last Muhammad, one of whose titles is "the seal of the Prophets." Moslems consider Muhammad to be the Messenger and Prophet of God and the most perfect of all people, but they nonetheless do not attribute any divinity to him. However, some mystics consider the Prophet to be saintlike and practically divine, a line of reasoning generally more pronounced among the Shi'a.

Another basic belief is in the Day of Judgment, when God will appear on his throne to judge the deeds of humanity. The Quran reminds the faithful that " . . . those who believe and who do good works and establish worship and pay the poor their due, their reward is with their Lord and there shall be no fear come upon them, neither shall they grieve." These beliefs enjoin a number of specific duties or obligations that are incumbent upon all adherents.

DUTIES AND RITUAL IN ISLAM

Islam tends to classify most human activities into two categories: those that are permissible or "lawful" (*halal*), and those that are forbidden (*ḥaram*). In a general sense this contrast can be likened to the distinction between acts that contribute to a state of spiritual grace or purity, as opposed to those that pollute or taint. The context of the act is all important in determining whether it is regarded as *ḥalal* or *ḥaram*. For example, an animal slaughtered in accordance with certain prescriptions is considered *halal* and fit for consumption; if not, it is *ḥaram* and should not be consumed by a believer. For the practicing Moslem, all actions should be performed in obedience to God's law as revealed to His Prophet Muhammad. This makes it almost meaningless to distinguish between the moral and legal aspects of an action; a sin is at once a crime.

Although no believer can completely achieve the full demands of the code of *ḥalal* in behavior, most Moslems display respect for it in their manner of dress, food handling, and general comportment. In Moslem countries, therefore, dress, the manner in which food is prepared and presented, attention to such details of physical appearance as beards, nails, and so on, all carry considerable symbolic importance. Minimally, they announce the membership of individuals in the Islamic community; their more careful observance signifies a deeper commitment. In this sense even daily and mundane activities take on the significance of ritual.

The most important duties of a believer are the following acts, which together express the Moslem creed. They are often referred to as the "Five Pillars of Islam":

> *1. The shahada, or profession of faith. As mentioned earlier, the recitation of the simple formula "I profess that there is no god but God; Muhammad is His messenger" constitutes the formal conversion of the reciter to Islam. The phrase usually constitutes the first words spoken into the ears of a newborn baby and should be the last on the lips of the dying.*

The simplicity of the *shahada* makes conversion an easy matter, a process that readily accommodates great variations in local custom and heritage. Dogma and catechism are secondary, and in many areas of the world new converts to Islam appear to know little about the faith beyond these few words. Once a convert to Islam, however, it becomes incumbent on the individual to learn its precepts and rituals; instruction thus usually follows rather than precedes the adoption of the faith.

> *2. The salat, or prayer enjoined on the Moslem five times a day: at dawn, midday, midafternoon, sunset, and nightfall. Worshippers face the direction of Mecca and go through the prescribed positions of the prayers, which they may perform anywhere after undergoing ritual ablutions.*

Many Moslems, especially younger men, do not observe the obligatory daily prayers. However, once an adult undertakes to pray daily, it is considered derelict for him or her to stop. Prayers should be offered wherever one finds oneself at the appropriate hours. Thus, intercity buses and trucks may stop by the roadside and passengers may get out to pray, with men and women in separate clusters. Farmers halt their plowing to pray in the fields. The ostentatiously pious may carry a prayer rug, but most simply place a clean handkerchief on the ground before them to which they touch their foreheads during the prayers. A number of men may pray together, but they do not coordinate their prayers unless they are in sufficient numbers to form a congregation with a prayer leader. The formal prayer, although ritually fixed, does not preclude the individual from offering personal beseechments or prayers of thanksgiving.

The main communal prayers take place at the mosque at midday on Friday. The men of a congregation stand in straight lines facing a semicircular recess called the *mihrab,* which indicates the direction of Mecca. The *imam,* or prayer leader, stands in front with his back to the group. At the time of Muhammad, women attended the Friday prayers but stood behind the men; later they prayed behind a screen. Today, women usually do not participate in the public prayers at all.

Many men who do not pray daily nevertheless attend regular Friday mosque services. In many small towns or villages not to attend would be to withdraw from the public life of the community. Although the mosque and its congregation are thought by many to exemplify the unity of Islam and equality before God, in practice the congregations of urban mosques, and sometimes even those of small towns and villages, reflect social and economic divisions within the society. In heterogeneous communities, members of different ethnic groups, tribes, and even occupations may well have their own mosques. Furthermore, within any one congregation, an implicit social hierarchy is expressed because men of prominence and power tend to pray in the front rows along with the learned of the community.

The Friday prayers are significant for other reasons. Following the prayers, the leading religious functionary or scholar present usually delivers a sermon, or *khutba,* which frequently goes beyond simple moral exhortation. The sermon is likely to deal with those topical issues that affect the community. At times sermons may also serve as the vehicles for political announcements; they may even call for insurrection against rulers perceived to be unjust. This has occurred frequently in history and quite recently in Iran, Syria, and Egypt, to name some prominent cases.

> *3. The zakat, or almsgiving. The Quran asks the believers to give alms as an expression of piety and as an aid to salvation. Almsgiving, which began as a voluntary act of piety and sharing within the small Moslem community, in time evolved into a legal duty, a tax levied on Moslems and administered by the state. Today, in most Moslem countries, the* zakat *has again become a voluntary act.*

Almsgiving, as *zakat* or otherwise, is an important means by which even those who do not closely adhere to other Islamic observances emphasize their identification with the community. A man gives small sums to beggers throughout the year, a woman gives a needy neighbor freshly baked bread, a family gives a sheep to an itinerant preacher, all in the spirit of *zakat.*

4. The sawm, or *fasting.* *Moslems are enjoined to fast the month of Ramadan, the month in which the Quran was first revealed to Muhammad; Ramadan is the ninth month of the Moslem lunar calendar. In the years when Ramadan falls in summer, its observance may entail a great deal of strain and self-discipline, especially in very hot regions like the Gulf area and Arabia. Even though some individual Moslems may choose not to fast, there exists strong public sentiment, even overt pressure, to observe Ramadan. The strength of this sentiment to conform publicly varies from one community to the other and from one country to the next. In Turkey restaurants remain open throughout Ramadan and many Turks continue to eat, drink, and smoke in public without fear of censure; in the Gulf states, all restaurants are closed.*

The fast is observed from dawn till sunset, during which time one may not eat or drink. Everyone is enjoined to observe the fast except children under the age of puberty. Pregnant women, the sick, military personnel on active duty, and those on a journey are exempt, but they should make up the missed days later. Those fasting, however, are free to eat anytime after sunset, and Ramadan nights usually turn into gay social occasions, with much visiting and exchanges of hospitality.

The end of the fast is celebrated by a feast, *'Id el-Futr,* which lasts for 3 days. The people celebrate by buying new clothes, by mutual entertaining, and by visiting their dead at the cemeteries. This latter is particularly important for women, who often visit cemeteries in groups.

Ramadan is also a time for reaffirmation of the faith; the Quran and religious sermons are broadcast daily over radio and television, and the people are exhorted to pray and to renew their faith. The shared experience of community members in observing the discipline of the fast further enhances the feeling of solidarity. Ramadan, as a time of renewal of the faith and as a uniquely Moslem celebration, acquires a special significance in countries under non-Moslem rule, as it did in Egypt, Syria, and North Africa. Under such circumstances, fasting becomes an expression of cultural pride, and implicitly a statement of opposition to foreign rule. When unpopular native secularist regimes are in power, Ramadan is similarly of potential political significance, as it proclaims at once Islamic unity and strength of belief, which could serve as vehicles for political opposition.

5. *The fifth pillar is the pilgrimage to Mecca, or* hajj, *which is required of every adult Moslem once in his or her lifetime, provided*

he or she is capable of doing so. It takes place during the first half of the twelfth lunar month, when pilgrims from all over the world converge on Mecca to join in performing the complex ritual that includes circumambulating the Ka'ba *seven times. Women undertake the pilgrimage provided they can be accompanied by their husbands or some other adult male who could serve as their protector. Under certain conditions, individuals may delegate a substitute to undertake the* hajj *for them.*

In the past, great caravans would form in Egypt, Syria, and Iraq for the difficult desert passage to Mecca.[12] Way stations stocked with food were placed along the routes, but the trip was often hazardous anyway. Today most pilgrims arrive by jet or bus, reside in clean accommodations, and benefit from the modern facilities provided by the Saudi government, which is responsible for handling as many as a million pilgrims a year.

Those who would be pilgrims temporarily withdraw from their own society and routine activities; they embark on a journey of the spirit as much as a trip abroad. Even before leaving home, they undergo purification and consecration as they suspend their everyday roles and acquire the special status of a departing pilgrim. The ceremonies that begin the *hajj* are performed in a personal state of ritual purity achieved by the observance of certain taboos and restraints, including sexual abstinence. The pilgrims wash ritually, are shaved, have their beards trimmed and nails cut. Each one then puts on a special robe which consists of two seamless white sheets. This simple garb is considered by Moslems today to exemplify the unity of Islam and the equality of all believers before God.

Upon entering Mecca, the pilgrims may take local guides who lead them through the ritual of the pilgrimage and otherwise assist them, particularly when the pilgrims know no Arabic. The first duty consists of circumambulating the shrine; later, following attendance at instructional sermons in the Great Mosque, pilgrims leave Mecca for Mount Arafat, where Muhammad received his prophecy. The next several days are filled with prescribed ceremonies, sermons, and prayers which recapitulate events in Muhammad's life.

The pilgrimage ritual culminates on the tenth day of the month, when the pilgrim sacrifices an animal (usually a sheep or a goat) which he or she had consecrated. This ceremony of the sacrifice is reenacted on the same day throughout the Moslem world as the head of each family

[12] For a fascinating personal account of a pilgrimage undertaken in the midnineteenth century by the famous British explorer-scholar Richard Burton, see his *A Personal Narrative of a Pilgrimage to al-Madinah and Meccah* (New York: Dover Books, 1964).

House exterior of an Egyptian *hajji* depicting events in his pilgrimage. (Photo courtesy of Ulku Bates)

sacrifices an animal. The feast is known in the Middle East as *Qurban Bayram,* or *'Id al-Kabir.* All who can afford an animal, including women who own property in their own name, are expected to make the sacrifice. Part of the meat is consumed in family feasting, but some is passed to the poor of the community. In this manner, even those who are not in the most sacred site of Islam participate in this major ritual associated with the annual *hajj.*

The *hajj* has always had practical economic significance, as did its pre-Islamic precedent, which was closely associated with trade. Traditionally, many pilgrims brought with them goods for trade, and a fair-like atmosphere attended the city of Mecca. Today, although few bring trade goods, many avail themselves of the numerous shops selling radios, cameras, jewelry, and all manner of imported luxury goods. More important, however, is the fact that the pilgrimage has always served to bring individual Moslems in touch with the fountainhead of their religion and to expose them to the theologians and savants who live and teach in Mecca. In the past and now, many African or Asian leaders, inspired by the *hajj* experience, have launched political careers as Islamic reformists.

Having fulfilled the required ritual in Mecca and its environs, the pilgrim returns home bearing a new, respected title, *hajji* (feminine: *haj-*

jiya). In Egypt a village *hajji* might well hire a painter to depict in colorful pictures on the white wall of his house the various places he visited, his mode of travel, and above all the Great Mosque and the Ka'ba. For some time after the pilgrimage, the *hajji's* house becomes the focal point of intensive visiting by friends and neighbors, all seeking to share in the *baraka*, or blessing of his experience. Pilgrims bring back prayer beads, perfumed oils, and mementos from the sacred city which they give out to visitors, especially those who had helped them prepare for the journey. Also, the *hajji* would most likely have brought back Japanese radios, watches, and cameras—a combination of trade and piety that is sometimes a cause for concern by officials in countries short of foreign currency. The average *hajji*, while enjoying a special status, soon returns to normal pursuits. For some, however, the *hajj* gives special impetus to already established careers in public life, whether in religious or in political arenas.

A lesser pilgrimage than the one to Mecca is to Jerusalem, site of the famous shrine-mosque, Dome of the Rock. The mosque was initially built by the caliph Omar to consecrate the spot from which it is believed that Muhammad took off on a nocturnal journey to Heaven. This legend is well known throughout the Islamic world and is the subject of poetry and art. One version has it that one night the Prophet was carried from Mecca to Jerusalem on a white-winged horse with a human face. In Jerusalem he saw Abraham, Moses, and Jesus at prayer together. Later he ascended the different heavens until he reached the seventh one, after which the horse took him back to Mecca. Sunni Moslems consider the Dome of the Rock to be their second holiest area after the *Ka'ba*.

As with Judaism, Islam imposes a number of dietary rules or taboos; among the foods that are considered taboo are pork, blood, and all alcoholic beverages. Forbidden also is the flesh of dead animals; only animals that have been ritually slaughtered by having their throats cut are considered fit for consumption.

The handling and serving of food takes on ritual significance in many circumstances. The believer is enjoined, for example, to treat bread, water, and salt with special respect and to avoid using the left hand for passing food or drawing from a common cooking pot or serving bowl; the left hand is associated with the performance of ablutions and unclean acts. Meals begin and end with words of praise to God, and the manner in which food is publicly offered, shared, and consumed has a significance that is more than simple etiquette.

Another ritual act that is not prescribed or even mentioned in the Quran but that is treated universally as a requirement of the faith is circumcision of males before puberty. This rite is carried out with as much public celebration and feasting as the family can afford. Circumcision, usually done between the ages of 4 and 7 constitutes a rite of passage; it marks the transition of the boy from the private domain of the household to the public one of the community. Simultaneously, it signals the separation of the young boy from his mother and his joining the world of the males. At this point he is likely to begin his formal religious education and is increasingly expected to identify with and observe the male codes of behavior. There is no equivalent rite of passage for the female child.

Duties to the dead are elaborate and closely prescribed by religious law and custom. Although practices vary from one locale to the other, certain rites are nearly universal. The body is washed by members of the same sex, shrouded in a single cloth, and interned by nightfall, if possible, and never later than the second day. Graves are dug so that the body can lie on its side facing Mecca, and care is taken that earth does not fall directly on the face of the individual. There is general agreement that the grave should be simple, and if adorned by a headstone or other marker, the ground immediately above the corpse should be left unobstructed. Under no circumstances are the dead to be brought inside the mosque, although the body may be carried into a mosque courtyard for a final prayer. A frequent sight alongside the inner walls of mosque courtyards is a wooded litter which serves to transport the dead to the graveyard.

The Islamic calendar is rich with days of feasting and ceremony. Even the weekly round of activities is punctuated by the Friday communal prayers and often, too, by activities that may precede and follow it. For example, Thursday evening is a time to visit the public bath, and in many communities women may keep a handful of token sweets they pass to children they encounter after their visit to the baths.

We could elaborate further on belief and ritual practice, but what we hope has emerged from our brief discussion of Islamic Middle Eastern society is a view in which Islam—the religion, its ritual, and its ceremony—is enmeshed in the daily activities of the individual. The way one holds one's hands while washing, styles of dress and hair, the manner of presenting food, the prayers on the lips of a traveler at the onset of a journey are all acts that weave together Islam into the living culture of the people.

Islam is often portrayed by its own scholar-jurists and by Western-trained Orientalists as a severely formal, even rigid legal tradition. Although Islam is the source of law and as such has generated a scholarly tradition emphasizing its jural relevance to virtually any situation, it is also much broader and richer. It is the basis for a moral order. It answers the question of what constitutes right and wrong, a moral person, a "good Moslem." Like all scriptural religions with codified ritual, it fundamentally distinguishes between the observant believer and the moral person whose life is informed by the ethical structure of Islam.

SOURCES OF ISLAMIC LAW

Having dealt with some of the basic beliefs of Islam and its most important ceremonial obligations, we now turn to a consideration of the sources of the system of beliefs, ritual, and duties that regulate a Moslem's relation to God and to fellow humans. There are basically two sources: the Quran and the *Hadith*, or Traditions of the Prophet. For Moslems, the Quran is the word of God; the term Quran means *recitation* and underscores the belief that it was revealed verbatim to Muhammad, who simply "recited" God's words: "The Koran (Quran) is the record of those formal utterances and discourses which Muhammad and his followers accepted as directly inspired. Moslem orthodoxy therefore regards them as the literal Word of God mediated through the Angel Gabriel."[13]

The Quran is believed to be a direct transcript of a tablet divinely inscribed and preserved in Heaven. Thus, as the literal word of God, the Quran may not be translated into any tongue other than Arabic and still retain the same validity. This explains the phenomenon observed in non-Arab Moslem countries, where some worshippers learn the Quran phonetically by heart and recite it without necessarily understanding the meaning of the words they utter. Considered holy and miraculous, the Quran is used by many as a talisman to ward off the Evil Eye. Verses from it are sealed in metal or leather containers and carried on the body to ward off evil and sickness and to ensure health and good luck.

Most of the Quran was committed to memory or written down on pieces of parchment or bone during Muhammad's lifetime. The text was collected in its entirety by the first caliph, Abu Bakr, but it was not until the third caliph, Othman, that a committee authorized a final version of the Quran. This standardized version is the only one that exists today.

The chapters of the Quran, known as *suras*, are arranged according to length by descending order, with the exception of the *fatiha*, which comes first. In all, there are 114 separate *suras*, the first half of which tend to be inspirational and exhortative in content and poetic in style; the latter half are more concerned with legislative and prescriptive matters. The Quran is written in beautiful, poetic, rhymed prose which,

[13] Gibb, *Mohammedanism*, p. 36.

when recited publicly, is intoned slowly in a melodic chant that is considered aesthetically pleasing to the listener. Special training is given in the art of Quranic recitation.

In addition to the Quran, Moslems are guided by the example of the life of Muhammad. The sayings and deeds of the Prophet form a system of social and legal usages collectively referred to as the *sunna* of Muhammad (course of conduct or path). The *sunna* is preserved in the form of short stories and anecdotes all dealing with what Muhammad said and did at various times. These anecdotes form an extensive literature in Islam known as the *Hadith*, sayings or Traditions.

The Quran and the *Hadith* are the major sources of the Islamic legal system known as *Shari'a*, or *Divine Law*. "As in other Semitic religions, law is thought of not as a product of human intelligence and adaptation to changing social needs and ideals, but of divine inspiration and immutable."[14] But because neither the Quran nor the Traditions provided a comprehensive and unified legal system, it was left for Moslem theologian/jurists, or *'ulama*, to interpret and elaborate the relevant texts and to construct the body of law, or *Shari'a*.

The *Shari'a* joins faith and practice as a comprehensive code establishing Islam as a way of life. The very centrality of the *Shari'a* to an Islamic moral and social order has challenged every secular political movement in the Middle East. It has raised problems for those who would legislate without reference to it; this is particularly true in the area of personal status or laws affecting marriage, divorce, and inheritance.

At the onset of Islamic rule, the *Shari'a* developed in a period of lively debate and discussion about the sources of law that would govern the rapidly expanding and heterogeneous Moslem community. The debates focused on the relative weight to be given to experience, rationality, and local custom. In time, four different schools of law crystallized within the dominant division of Islam known as the *Sunni*; each of these schools is considered "orthodox" by the others. The *Shi'a*, the other major division in Islam, have their own interpretations.

The four *Sunni* schools, or *madhhabs*, do not constitute separate sects, as they are in agreement in matters of doctrine and creed. Each was named after its jurist-founder and predominates in a different region. The Hanafis are found primarily in areas formerly governed by the Ottomans—Turkey, Iraq, Syria, Lebanon, and Lower Egypt—as well as parts of central Asia and India. The Maliki predominate in north and west Africa, upper Egypt, and the Sudan. The Shafi'i are represented in

[14] Ibid., p. 73.

parts of Syria, Iraq, and Turkey, but especially in Indonesia. The Hanbalis, who predominate in Arabia, are considered the most conservative in that they allow little scope for the use of reason or local custom.

Although purporting to be the universal basis for law, the *Shari'a* in practice today is evident primarily in matters pertaining to personal status.[15] These are cases that we might think of as falling within the jurisdiction of family and probate courts. Only Turkey has altogether rejected the *Shari'a* as a basis for national legal codes. Even in that country, there is a growing movement instigated by the religiously motivated far right for the return of a *Shari'a*-guided court system. Elsewhere, the *Shari'a* serves to adjudicate matters of marriage, divorce, child custody, inheritance, adoption, and public decorum. In most countries commercial and criminal legislation tacitly ignores the *Shari'a*, and such codes are derived from European sources. Two major exceptions are Iran and Saudi Arabia, where, in principle at least, the *Shari'a* is adhered to in both civil and criminal matters. Today, the single most important controversy concerning the *Shari'a* throughout the Middle East is its relationship to the rapidly changing relative status of the sexes and to changing patterns of family life, topics we take up in later chapters.

RELIGIOUS LEADERSHIP: THE *'ULAMA*

Our cursory discussion of the *Shari'a* and its central place in Islam might suggest that the ideal Moslem state or society would be a theocracy. This, however, would be an inaccurate conclusion. One paradox of *Sunni* Islam is that its great emphasis on the all-pervasive scope of Islamic law disavows any fundamental distinction between religious and nonreligious domains. What a Christian might consider secular and nonsecular domains of life are undistinguishable in Islamic thought; from this perspective, the terms "secular" and "theocratic" become meaningless. In fact, if one were to use them to describe the ideal Islamic society, it would have to be as a "lay theocracy." For even though all legitimate power belongs to God, since Muhammad's death He has had no earthly spokesperson, nor is He served by a special caste or priesthood. In principle, each Moslem is equally capable of communicating with God without any mediation, and each may aspire to any position within the community. The *Shi'a* situation is rather different, and we take it up in the following chapter. For *Sunni* Middle East society, the many distinc-

[15] See Majid Khadduri and Herbert Liebesny, eds., *Law in the Middle East* (Washington, D.C.: The Middle East Institute, 1955); also Reuben Levy, *Social Structure of Islam* (Cambridge, England: Cambridge University Press, 1957).

tions of religious rank, learning, and even institutional authority do not form one unified hierarchy, let alone a church. The most exalted religious office, for example the Sheikh al-Azhar in Cairo, head of the most prestigious mosque-university of *Sunni* Islam, has no legitimate authority over the humblest village sheikh. In the absence of a church, the definition of "orthodoxy" becomes problematic.

This situation is complicated by the fact the the *Shari'a* is, again in principle, considered to be immutable, and there is no final arbiter of its content. This results in another paradox—while viewed as eternal and fixed, the *Shari'a* is also everchanging. Within this contradiction lies fertile ground for the development of reformist ideologies, mystical movements, and schismatic rebellions, all of which have been part of Islamic history.

Given the absence of church and priesthood, how is religious life structured within the society? Who instructs the young, interprets the *Shari'a* and leads congregations? And how does one account for the obvious disparities in rank and influence among the different religious personages?

Until quite recently, as in medieval Europe, all formal learning was religious. Learned individuals, or the *'ulama,* were by definition religious scholars—whether they were concerned with the Quran and its exegesis or with astronomy or medicine. They could be self-taught or more likely the student-disciples of one or another well-established scholar. With time, the *'ulama* came to form a special group, with its own insignia of distinctive turbans and robes and with more or less agreed-upon ranking procedures and rules for recruitment. Status differentiation was based on a combination of factors, which included scholarly achievement, the personality of the individual, peer recognition, and general support of the local community. Particular scholars emerged as authorities in particular areas, acquired vast reputations, and attracted wide followings. The lack of a formal hierarchy is reflected in the terminology where the same term, *sheikh,* may be used for *'ulama* of all levels. This is in contrast to the Shi'a practice in which clerics are more carefully distinguished by rank.[16]

The *'ulama* as a group trained the teachers, preachers, and bureaucrats of the society, and in some periods they themselves con-

[16]The *'ulama* have constituted an important topic of scholarship. For a convenient introduction, see the volume edited by Nikki Keddie, *Scholars, Saints and Sufis: Muslim Religious Institutions since 1500* (Berkeley: University of California Press, 1972). The volume is divided into two sections: The first deals with history and organization of *'ulama,* the second with popular religious institutions. For a recent anthropological study of the *'ulama* in Iran, see Michael M. J. Fischer, *Iran: From Religious Dispute to Revolution* (Cambridge and London: Harvard University Press, 1980).

stituted a powerful patrician class. Their students spent variable periods of time in places of learning, or *madrasas,* where the individual would study under the tutelage of a particular master. Those who were successful themselves became the future *'ulama*—theologians-cum-jurists, judges, government advisors, and even ministers. The *'ulama* were, and still are, the self-appointed guardians of the *Shari'a* and its executors within the society. It was they who ultimately legitimized the ruler and his actions to the population—a role that conferred considerable power on them, even if indirectly. In some countries, they were incorporated into state bureaucracies, primarily as *qadis,* or judges, and school administrators.

Traditionally, the *'ulama* derived their income and considerable power as administrators and beneficiaries of trusts in land and urban property (called *waqf*) set up by wealthy donors. In some areas a sizable portion of the arable land was *waqf,* and the great mosque complexes of major cities were supported not only by the donations of the faithful, but by the income derived from renting the many shops and other property they owned in the city.

The advent of European-style education and the expansion of scientific curriculum everywhere in the nineteenth century presented the *'ulama* with a serious challenge to their historic monopoly on education. Today the term *'ulama* refers exclusively to those trained in the religious tradition, and recruitment to this group in most countries reflects the 'ulama's increasingly restricted role in modern society. The governing elite and the well-to-do almost inevitably educate their children in Western-style schools and universities. Quranic schools today are primarily attended by the very young, who come to learn the rudiments of religion and to memorize parts of the Quran. Virtually every large village in the Middle East aspires to have its own Quranic school supported by the community, and failing that, families get together to hire a religious teacher (*mullah, fqih,* or *hoja*) for their children. This, however, is usually in addition to secular education.

The relationship between the *'ulama* and the state, both in Sunni and Shi'a Islam, is much too complex a subject and is outside the scope of our presentation. Suffice it to say here that generally speaking, in Sunni countries the *'ulama* have tended to work closely with the ruling establishment. Under the Ottomans, for example, the sultan came to appoint or ratify major religious posts and even attempted to centralize and control religious leadership through the creation of the post of *sheikh-al-Islam,* or paramount sheikh. In Iran, the Shi'a *'ulama's* position has generally been, at least since the eighteenth century, in opposition to the state. Following the revolution of 1979, however, the *'ulama* in Iran have come to constitute a major locus of power and authority in the state.

So far, our emphasis has been on the origin and development of Islam, its basic tenets and rules. In the following chapter, we take up the doctrinal and historical bases for the important sectarian divisions, including Shi'ism, as well as the development of mysticism in Islam and its different popular expressions today.

CHAPTER THREE
ISLAM AS CULTURE, ISLAM AS POLITICS

The Prophet's vision of a community of believers united in their common faith in one God was soon embodied in a political form, the Islamic state. The unity of this polity was, however, more apparent than real, masking as it did great social and economic tensions and fundamental disagreements. In short, the Islamic community, like any religious or ideological movement, had to accommodate social and cultural reality. The early community had, in fact, brought together quite disparate tribes from northern and southern Arabia—people pursuing different ways of life and speaking different dialects, but all seeking expression and even contending for power within the nascent Moslem state. Immediately upon Muhammad's death, the community was torn by a number of political challenges, including the secession of some tribes who attempted to follow other, so-called "false Prophets." The first caliphs, however, succeeded in keeping the community together and in forging a vigorously expanding empire. The borders of this empire soon reached the Pyrenees in the west and Afghanistan in the east. The conquest of this vast area, much greater than that conquered by Alexander, was rapidly accompanied by the conversion of its diverse peoples to Islam, a fact that must attest to the power of the new Islamic order.

Only the heartland of Byzantium in Asia Minor and its capital Constantinople withstood the forces of the early Arab armies. The great Sassanian empire of Persia disintegrated rapidly as its armies were defeated and its capital, Ctesiphon, destroyed. Perhaps it is because of the Islamic movement's great and rapid success that correspondingly powerful centrifugal forces soon came to threaten it. Within a quarter of a century following Muhammad's death, dissenting political movements using the idiom of Islam had broken away from the main body of Moslems. Thereafter, throughout Middle Eastern history, Islam as a revolutionary force counterbalanced Islam as an established political and social order.

Islamic movements from the earliest periods have had their origin in political protests whose legitimacy was always expressed in a religious idiom. This is consistent with the view that permits no theoretical distinction between religion and politics. Thus any political challenge to the state almost invariably assumes religious overtones, just as any doctrinal disagreement can easily imply a potential political threat.

SCHISMS IN ISLAM: THE SHI'A

The single most important schismatic movement in Islam is that of Shi'ism, whose adherents predominate today in Iran, southern Iraq, and in parts of Turkey, Syria, and Lebanon.[1] Until recently, in Syria and Lebanon the rural Shi'a communities constituted a poor and depressed population, dominated in feudal fashion by large landlords who included Sunni and Christian as well as Shi'a families.

As we described earlier, the Shi'a split arose from a dispute over who should succeed the Prophet after his death in A.D. 632. Those who championed 'Ali in effect restricted the legitimate succession to a direct line of descent from Muhammad through 'Ali, his patrilineal cousin and son-in-law. The majority position prevailed in choosing the successor from among the Prophet's companions. Although 'Ali did, in time, become the fourth caliph, he immediately met opposition from an already entrenched Moslem leadership—most importantly, from the governor of Syria, who was a member of the powerful Ummayad clan. The civil war that followed rent the Moslem community. 'Ali was ultimately defeated and killed. In many respects the successful opposition to 'Ali on the part of the Ummayads reflected the fact that the center of

[1] For a general introduction to the history of the Shi'a movement, see Dwight M. Donaldson, *The Shi'ite Religion, A History of Islam in Persia and Iraq* (London: Luzac, 1933).

political power had already shifted out of Arabia to the ancient lands of the Mediterranean. From then on Mecca ceased to be a center of political power, a role assumed by Damascus and later Baghdad.

The civil war in which 'Ali was defeated had many important consequences for the subsequent history of the Middle East. Quite apart from the emergence of Shi'ism as a political movement, it also affected the geographic and ethnic distribution of the population. For example, a large number of the supporters of 'Ali were Yemeni tribespeople who came to settle in the region of Kufa in southern Iraq. The Yemenis considered themselves culturally superior to the northern Arabian tribes and felt that they had been politically shunted aside by the ruling Ummayad aristocracy. They were among the first to rally to 'Ali's party, thus marking the beginning of a significant Shi'a presence in Iraq.

Another sectarian movement founded at this time was that of the Kharijites, or "seceders," who withdrew from 'Ali's camp during his war with the Ummayads. The Kharijite movement acquired considerable significance as the vehicle for local rebellions in Syria and Iraq, but its political importance declined following the eighth century. Descendants of the Kharijites are found today in Oman, Zanzibar, and parts of east and north Africa, where they are known as 'Ibadis. They represent a moderate branch of the original movement which espoused the simplicity and democracy of the original Moslem community at Medina. In Oman the 'Ibadis constitute the ruling oligarchy, as they did in Zanzibar until the midfifties. The Kharijites established a precedent that has come to characterize much of Islamic dissidence. By defining those Moslems who do not espouse their beliefs as "apostates," they challenged the very definition of what constitutes an Islamic community and redefined its boundaries in ways that emphasized already-present ethnic divisions.

Losing out to the Ummayads and persecuted throughout the Islamic world, the Shi'a were soon forced to go underground. Secret but active, they began to rally dissidents with a proselytizing zeal that made them grow in number and influence. Increasingly, segments of the non-Arab populations of Iraq, Syria, and Persia joined their ranks. Historians feel that the Shi'a movement represented a populist ethnic reaction to the exclusiveness and discrimination practiced by the conquering Arab tribes, who tended to form a privileged caste of warriors ruling over the indigenous people. Even though the latter had converted in large numbers and were, at least in principle, the equal of the Arabs, the social and political distinctions between the Arab conquerers and the converted population remained important. Quite apart from being a vehicle for political dissent, Shi'ism also allowed the perpetuation of local beliefs and practices that were contrary to Sunni formulations. Shi'ism flourished as a folk or popular cult, much as do the mystic orders and cults today, which, we might add, are still often viewed as oppositional to the establishment.

Although a Shi'a dynasty was to achieve power in Egypt during the tenth and twelfth centuries, the most significant and lasting political success of the Shi'a came in 1502, when Shah Ismail seized the throne of Persia and made Shi'ism the state religion. The formal attainment of power in Persia after a long period of struggle further shaped the dogma and organization of Shi'a Islam in ways quite distinct from that of Sunni Islam. We now treat these in turn.

SHI'A BELIEFS: A CULTURE OF DISSENT

The Shi'a hold that the family of Muhammad, and specifically 'Ali and his male descendants from his marriage to Fatima, the daughter of the Prophet, possess supernatural powers and are uniquely qualified for the caliphate and the leadership of the Islamic community. Thus, the first three caliphs and all succeeding ones are really usurpers, having displaced the only legitimate successor to Muhammad—namely 'Ali, who is the first *Imam*. Whereas to the Sunni, the term *imam* simply designates the leader of the Friday mosque prayers, to the Shi'a it has a very special meaning. In fact, for the Shi'a, belief in the *imam* constitutes the sixth tenent or pillar of religion, in addition to the five discussed in the previous chapter.

The Shi'a doctrine of the imamate holds that the imam is the agent of Divine Illumination and the medium of Divine Revelation. Because of this, Shi'a imams are considered to be sinless and infallible, and are thus empowered with great authority to pronounce dogma. This authority, plus a body of "esoteric knowledge" which is believed to have been originally bestowed on 'Ali, have since been passed down to a select number of his male descendants. The various Shi'a sects differ over the question of the order of succession to the imamate, some championing one over another of the various descendants of 'Ali.

This doctrine that regards 'Ali and his descendants as special beings, endowed with supernatural powers and repositories of Divine Truth, is in sharp contrast with the insistence of mainstream Sunnism on the humanity of Muhammad, the Messenger, and his role as the last of the Prophets. It further makes of all descendants of 'Ali a caste-like group who even today enjoy a special status among the Shi'a, where they are collectively known as Sayyids.

The Sayyids marry among themselves and are subject to special consideration in that violence should not be directed against their person or property; to do so, even inadvertently, is considered a sacrilege. Some Sayyids are thought of as holy. The pious may give the Sayyid money as a tithe. Even in Sunni communities, descendants of the Prophet's family are accorded special recognition. It should be emphasized here that the status of Sayyid does not in itself confer wealth, power, or even significant social influence on the individual. In some communities, the Sayyids may form a loosely defined group of landless and socially marginal mendicants sustained by charity. Elsewhere, for example, in a region in southern Arabia, until recently, the Sayyids as a group constituted a wealthy, ruling oligarchy who jealously preserved their control over property and power at the same time that they perpetuated their genealogical claim to holy status.[2]

Another distinguishing theme in Shi'ism is that of martyrdom or Passion, centered on the figure of the second imam, Hussein. 'Ali was succeeded to the imamate by his son Hussein, who, along with members of his family, was ambushed and massacred by Ummayad troops near the city of Karbala in southern Iraq in A.D. 671.[3] This tragedy, following the assassination of 'Ali by a Kharijite fanatic, has given Shi'a Islam a specifically tragic cast and a definite proclivity towards a cult of martyrdom. The devotion of the Shi'a to the tragic figure of Hussein and his descendants, many of whom met with violent deaths, adds a dimension to Shi'ism not found in Sunni belief.

The betrayal and martyrdom of Imam Hussein is dramatically enacted yearly during the month of Muharram with rites that portray the tragedy with much emotion and grief. The Muharram rites themselves vary greatly among the Shi'a communities and even within Iran. Probably no other public display so impresses the non-Shi'a observer as does the intensity of feeling expressed in the processions and Passion plays, which culminate on the tenth day of Muharram. The murders of Hussein and family members are then enacted, often with local Sayyids playing leading roles in dramatizations in which the severed head of the murdered Hussein addresses the community. Public parades and demonstrations of grief are joined in by spectators who sometimes engage in violent self-flagellation. The predominant message concerns the injustice of this world, oppression, and the ultimate triumph of truth and justice. The plays dramatically portray the perfidy and oppression of rulers, especially Arab and Turkish caliphs. Christians and Jews are frequently represented as collaborators with the oppressors. Unpopular current leaders and governments may be reviled by being associated in the plays with the foes of Hussein.[4]

[2] A. Bujra, *The Politics of Stratification: A Study of Political Change in a South Arabian Town* (Oxford: Clarendon Press, 1971).

[3] For a brief period, another son, Hassan, was designated the imam after the assassination of 'Ali.

[4] See Gustav Thaiss, *The Drama of Husain*, Ph.D. dissertation, University of Washington, 1973; also his article, "Religious Symbolism and Social Change: The Drama of Husain," in

The tomb of Hussein in Karbala and that of 'Ali in Najaf, like those of subsequent imams located in Baghdad, Qum, Meshhed, and elsewhere, are venerated as shrines and visited by pilgrims. The sanctuaries of Karbala and Najaf are considered holy ground, and the pious make efforts to bring the bodies of their dead for burial in their environs. Pilgrims still bring donations of gold, fine jewelry, carpets, and other valuables to the major shrines, which historically have been endowed with great wealth. Shahs, sultans, emperors, and other rulers have all made gifts in demonstration of piety and perhaps in pursuit of legitimacy. Even quite recently the former Empress of Iran, Farah Pahlavi, donated two massive gold and jewel-encrusted doors to the sanctuary of Karbala. The cupolas of leading shrines are leafed in gold, and colored mosaics, mirrors, and gold and silver calligraphy embellish the walls.

The effect on the pilgrim must be quite overwhelming, for, as the German nineteenth century traveler-scholar Nöldeke writes for the sanctuary of Hussein in Karbala:

> *. . . the general impression made by the interior must be called fairy-like, when in the dusk—even in the daytime it is dim inside—the light of innumerable lamps and candles around the silver shrine, reflected a thousand and again a thousand times from the innumerable small crystal facets produces a charming effect beyond the dreams of imagination. In the roof of the dome, the light loses its strength; only here and there a few crystal surfaces gleam like the stars in the sky.*[5]

The majority of the Shi'a today belong to the Imami, or Twelvers sect. They believe in an unbroken chain of twelve imams, beginning with 'Ali and ending with Imam Muhammad al-Mahdi, who, while still a child, disappeared in the mosque of Samarra in Iraq in A.D. 874. He was rumored to have been secretly murdered by the 'Abbasids, the dynasty then in power. The Twelveth Imam is believed to be simply hidden in a state of *occultation*, or noncorporeal existence, and one day he is expected to return as the *Mahdi*, or rightly guided one, to usher in a reign of justice and peace in a world full of sin and injustice. The Shi'a thus share a messianic concept with the Jews and the Christians, a concept that also seems to have its antecedents in the early Zoroastrian religion of Persia.

Nikki Keddie, ed., *Scholars, Saints and Sufis: Muslim Religious Instructions since 1500* (Berkeley: University of California Press, 1972), pp. 349–66.

[5] A. Nöldeke quoted in the *Shorter Encyclopedia of Islam*, eds. H. A. R. Gibb and H. H. Kramers (Ithaca: Cornell University Press, 1955), pp. 360–61.

The cult of martyrdom and the messianic belief in the return of the Hidden Imam are at the core of the Shi'a ethos.

The concept of the *Mahdi* has its echo in a widespread folk belief among Sunni people that a *Mahdi*-like leader will emerge to restore "True Islam" and obliterate injustice and oppression. This is a variant on what anthropologists call *millenarianism,* a belief in the restoration of an ideal but lost order. Millenarian movements are often triggered by severe cultural dislocations, such as those occasioned by foreign invasion or colonialism. One such Islamic case was the *Mahdi* revolt in the Sudan in the late nineteenth century, which protested Anglo-Egyptian rule. Although put down after bloody fighting, the *Mahdiyya* movement marked the beginning of Sudanese nationalism.

THE SHI'A CLERGY

In comparison with their Sunni counterparts, Shi'a clergy are more hierarchical and centralized in their organization; historically, they have also been much more prone to take oppositional roles in their relationship to government. Their claim to authority derives from their role as the deputies of the absent imam. In an influential article on the role of Iranian *'ulama* as opposition leaders, Hamid Algar quite prophetically foresaw their importance in rallying the populace against the unpopular regime of the late Shah of Iran and even predicted the leadership of the Ayatollah Khomeini.[6] The very ideology of the imamate, with its emphasis on the return of the imam, renders all government, even those formally affiliated with Shi'ism, as usurpatory. For generations the *'ulama* have reacted against the tyranny of autocratic rulers, using the martyrdom of Hussein as exemplification of sacrifice in the attainment of temporal justice.

Higher levels of clergy, which in Shi'ism are distinguished by clear ranks, have propagated their role as arbitors of the imam's authority and interpreters of his will. Those holding the title of *mujtahid*—that is, those capable of interpreting the law—serve as community leaders. As spiritual teachers, they, and especially the high-ranking *mujtahids,* called *ayatollahs,* offer themselves as moral guides. In theory, each individual Shi'a selects a particular *mujtahid* to emulate and to recognize as his or her personal authority on moral issues. This entails the submission of the layperson to the opinions of the spiritual guide. Because *muj-*

[6] Hamid Algar, "The Oppositional Role of the Ulama in Twentieth Century Iran," in Nikki Keddie, ed., *Scholars, Saints and Sufis,* pp. 211–31.

Mass prayer in Tehran led by a Shi'a *mujtahid.* (Photo courtesy of the United Nations/ John Isaac)

tahids can be called upon to advise and make judgments on all aspects of life, their pronouncements are of obvious political significance.[7]

Both the *mujtahids* and lower-ranking clerical functionaries called *mullahs* are in close touch with the masses as they are called upon to offer very personalized guidance, a role not usually played by Sunni *'ulama.* Still, some men, often called *sheikhs,* may less formally fill a similar role, albeit one not endorsed by orthodox doctrine.

In Shi'a communities men, women, the sick, the healthy, the rich, and the poor alike regularly seek out their spiritual mentors with gifts, attend their prayer sessions, read their religious texts, and call upon them for solace when faced with misfortune. People gather in the presence of important *mujtahids* bearing petitions for help in dealing with government authorities, to get sons out of jail, secure jobs, rectify

[7] Two studies of the political role of the Shi'a*'ulama* in Iran, one of the nineteenth century and one for the current period, are Hamid Algar, *Religion and State in Iran, 1785–1906* (Berkeley: University of California Press, 1969) and Sharough Akhavi, *Religion and Politics in Contemporary Iran* (Albany: State University of New York Press, 1980). Also Nikki Keddie's, *Sayyid Jamal al-Din al-Afghani: A Political Biography* (Berkeley: University of California Press, 1972) is a very good account of an anti-Western Islamic reformer in the late nineteenth century.

bureaucratic abuse, and the like, thus reinforcing the prestige and influence of the clerics.

In Iran donations by the pious over generations have created vast holdings of property set aside for the support of shrines, clerics, mosques, and religious schools. Moreover, followers give directly to their chosen mentors. This wealth has tradionally given the *'ulama* an independent base. Both Pahlavi Shahs attempted to seize those holdings and break the power of the *'ulama*. Each in his turn failed.

The Iranian *'ulama* perpetuate themselves much like their Sunni counterparts through their *madrasas* (school-seminary), to which students are drawn from all over the country. Students without means are supported through endowments of the school-seminary. When deemed ready, they go back to their communities; a few select may stay to pursue further study and to themselves become *mujtahids*. Both the cities of Najaf in Iraq and Qum in Iran are considered to be major centers of Shi'a learning; advanced scholars spend time in their schools and libraries.[8]

SECTS WITHIN SHI'ISM

Shi'ism itself is fragmented in a way not encountered in the Sunni world. A number of Shi'a sects disagree so profoundly on central issues of dogma that they scarcely recognize the legitimacy of each other's contrasting views. The dominant schismatic divisions have to do with the nature of the imamate and the order of succession. Divisions within Shi'ism are further complicated by a principle they all share, that of "dissimulation of belief." This pragmatic approach allows a person to dissimulate or conceal his or her real belief when threatened or in a hostile environment. This practice has served to shield a proliferation of divergent rites and beliefs, even among sects that share the same name, and it raises questions about the definition of "orthodoxy" in a religion that lacks a church.

Four Shi'a or Shi'a-derived groups deserve mention here because they illustrate some of the different directions taken by Shi'ism, as well as being of ethnographic importance in their own right. The four are the Isma'ilis (or Seveners, as they are sometimes called), the Nusairis (or Alawis), the Druze, and the Zaidis.

The Isma'ilis, who are perhaps the least important today in the Middle East, go back in their origin to A.D. 765, when they broke off with

[8] For an excellent account of the Shi'a madrasas or seminaries of Qum and the culture of religious learning in Iran, see the recent study by Michael M. J. Fisher, *Iran: From Religious Dissent to Revolution* (Cambridge, Mass.: Harvard University Press, 1980).

the Imamis over the choice of a successor to the sixth imam. For them, the seventh imam, or Isma'il, is their last acknowledged imam and the one whose reappearance as the *Mahdi* is awaited. In the wake of local revolts, early Isma'ili states were founded in Syria and Bahrain during the tenth century. But the apogee of their political success was reached in Egypt with the accession of the Fatimid dynasty to power during the tenth and eleventh centuries. Later, in Persia, a strong and well-organized community of Isma'ilis was established around the mountain fortress of Alamut near Qazvin. From their mountain stronghold, emissaries were sent out to perpetrate acts of political sabotage and assassination attempts against the rulers. As the sect was reputed to use hashish in their ritual, the Isma'ilis of Alamut came to be known as *al-Hashashin* (hashish smokers), whence came the English word *"assassin."*

The Isma'ili stronghold in Persia was destroyed in 1250 by the Mongols, after which they ceased to be politically important in the Middle East. Today, Isma'ilis are found in India, Syria, the Persian Gulf area, and in parts of East Africa. A group among them, the Nizaris, pay special homage to the Agha Khan, whose family claim descent from the Seventh Imam. What distinguishes the Isma'ilis even today is their insistence on an allegorical and esoteric interpretation of the Quran. This practice separates the religious initiate from the layperson in a manner not found among the majority of Moslems. The paradox in the case of the Isma'ilis is clear. What began as a populist, near-revolutionary movement against the Sunni oligarchy and its ruling institutions, has, in time, engendered an ideology that sustains a favored few in positions of leadership and mediation with God.

An offshoot of the Isma'ili movement is the Druze sect, whose origins go back to the eleventh century, when a Persian named al-Darazi began to preach an Isma'ili doctrine among the mountain-dwelling rural populations of southern Lebanon and parts of Syria. Today the Druze are divided among the states of Syria, Lebanon, and Israel.

Another Shi'a-derived sect is that of the Nusairis, also known as Alawis, whose members are found today in northern Syria and parts of Turkey and Iraq. Numbering about 300,000, Alawis are largely rural. They practice a religion carefully concealed from outsiders, with doctrines and rituals founded on an esoteric interpretation of the Quran. The Alawis, long a denigrated and weak rural minority in Syria, today number among their members a majority of the political leaders who owe their initial power to their role in the Syrian army, to which they were recruited by the French. In Turkey the Alawi situation is rather different. Long regarded with hostility by the Sunni majority, they only recently have begun to mobilize politically. Frequently, they are found in one or the other of the leftist parties and have been involved in armed communal conflict in at least two major towns in the East—Sivas and Marash. In Syria and in Turkey the emergence of the Alawis on the regional and national political scene has occasioned a powerful backlash on the part of the Sunni majority.

The Zaidis, who are primarily found in Yemen today, represent an interesting development within Shi'ism. They come closest to Sunni orthodoxy in that they have living imams who, in principle, are chosen by the community. The Zaidi imams ruled Yemen from the late ninth century until the 1960's. Even today, the Zaidis, a numerical minority, tend to dominate the political and economic scene in North Yemen. We thus see in both the cases of Oman and until recently in Yemen a small oligarchy using the claim of holy descent and religious leadership as a source for political authority and legitimacy in ruling an ethnically and tribally fragmented population.

ISLAMIC MYSTICISM: THE SUFI WAY

The texture of Islamic religious experience is further enriched by mysticism. Middle Eastern people from all regions, rural and urban alike, have from the beginning of Islam been heirs to a long tradition of mysticism and asceticism in which personal piety and emotional catharsis go together. Sufism, as Islamic mysticism is commonly called, is a word that embraces a vast array of beliefs, rituals, and even formal institutions, such as orders and brotherhoods.[9]

In many respects, Sufism represents the side of Islam that is eternally changing, responsive to the moods and exigencies of the moment. Through its openness and individuated character, Sufism has buffered Sunni and Shi'a orthodoxies from many of the challenges and pressures that might have engendered a major reformation. Many of the recurring revitalization and reform movements within Islam have sprung from Sufism and have expressed themselves in religious orders and brotherhoods that parallel, and sometimes even challenge, but do not replace, the structure of the formal religion. The significance of Sufism for facilitating the spread of Islam in India, central Asia, and Africa cannot be underestimated. It is everywhere a vehicle for popular local beliefs and practice, and as such Sufism tends to be expressed differently in different areas. The political role of organized Sufism remains powerful as a potential mobilizer of mass sentiment. Sufism therefore

[9]For a good general introduction to Islamic mysticism, see Reynold A. Nicholson, *The Mystics of Islam* (London: Routledge and Kegan Paul, 1963), and J. Spencer Trimingham, *The Sufi Orders in Islam* (Oxford: Oxford University Press, 1971); see also Annamarie Schimmel, *Mystical Dimensions of Islam* (Chapel Hill: University of North Carolina Press, 1975).

has both its private dimension, rooted in intimate individual religious experience, as well as its public dimension in organizing politically significant movements.

The origin of the term *Sufi* is obscure. The assumption is that it comes from the garments of rough, undyed wool (or *suf*) worn by the early mystics in Baghdad. As early as the eighth century, these men wandered from town to town preaching asceticism, spiritual discipline, and ecstatic communion with God. The Sufis emphasized emotional spontaneity and sought to free the religious experience from the legalistic demands of the *Shari'a*. By allowing each individual to seek his or her own spiritual path, Sufism came to represent a popular reaction to the increasingly rigid religious establishment.

In Sufism, the ultimate goal is loss of individual consciousness and complete union with Truth or "Ultimate Reality." This state, mystics believe, may be achieved through intuition and not through the exercise of reason. Mystics seek to achieve their goals by renouncing worldly concerns and through various spiritual and physical exercises. The Sufi path to God, or *tariqa*, consists of a number of stages through which initiates have to pass. On the way, they are assisted by personal teacher–masters, their sheikhs, or *pirs*. Once the initiates successfully pass through all the stages, they will attain their ultimate goal—a new kind of consciousness consisting of *ma'rifa*, or knowledge, and *haqiqa*, or truth.

This mystical progress is described in a Turkish saying:

> To know *Shari'a* is to know that yours is yours,
> and mine is mine.
> And to know the *tariqa* is to know that yours is yours,
> and mine is yours, too.
> But to know the *ma'rifa* is to know that there is
> neither mine nor yours.

From its inception and throughout its history, poetry was the most important medium for the expression of Sufi sentiment, and many of the leading mystics were also celebrated poets. Al-Hallaj, the ninth-century Sufi martyr, was born in Persia but spent most of his adult life teaching and writing in Baghdad. A great poet and a persuasive teacher, Al-Hallaj gathered a large number of followers and grew so influential that he posed a threat to the established authority of the *'ulama*, who eventually conspired with the ruling dynasty to have him executed in A.D. 922. The charge against him was heresy. In what is perhaps the most celebrated statement of early Sufism, Al-Hallaj had declared, "I am the Absolute Truth."

In his poetry, Al-Hallaj celebrated mystic love and the harmony that follows the complete union with God:

> I am he who I love, and he whom I love is I,
> We are two spirits dwelling in one body.
> If thou seest me, thou seest Him,
> And if thou seest Him, thou seest us both.[10]

Even before Al-Hallaj, the theme of unbound, unconditional love had become an integral part of Sufi belief. Rabi'a Al'Adawiya (d. 801), a freed female slave who lived in Iraq, taught that love should replace fear as the motive for religious devotion. The story is told that once Rabi'a was seen walking in the streets with a torch in one hand and a pitcher in the other. When questioned, she answered: "I want to throw fire into Paradise, and pour water into Hell, so that these two veils may disappear and it becomes clear who worships God out of love, not out of fear of Hell or hope for Paradise."[11]

One of the most popular folk characters in the Middle East, Mulla Nasrudin, is also a vehicle of Sufi teaching. The stories of Mulla Nasrudin, told in coffeehouses and village squares as jokes or moral fables, form a distinct literary genre specifically used by Sufis to express the subtlety of mystical knowledge, as the seemingly simple tales can be interpreted at many levels. The following two examples, chosen from a vast repertoire, serve to illustrate a style of Sufi teaching and at the same time introduce one of the most famous figures of popular culture in the Middle East. They are taken from the work of Idris Shah, the Pakistani Sufi who is largely responsible for the current popularity of Sufism in the West:

> *Mulla Nasrudin was walking along an alleyway one day when a man fell from a roof and landed on top of him. The other man was unhurt—but the Mulla was taken to the hospital.*
>
> *"What teaching do you infer from this event, Master?" one of his disciples asked him.*
>
> *"Avoid belief in inevitability, even if cause and effect seem inevitable! Shun theoretical questions like: 'If a man falls off a roof will his neck be broken?' He fell—but my neck is broken."*[12]

As Shah explains, the preceding tale is used by Sufi masters to teach the initiate to question belief in simple cause and effect. The

[10] Louis Massignon, *La Passion d'al Hosayn ibn Mansour al-Hallaj, Martyr Mystique de l'Islam*, Vol. II (Paris: P. Geuthner, 1922), p. 518.

[11] Schimmel, *Mystical Dimensions*, p. 98

[12] Idris Shah, *The Sufis* (New York: Doubleday, 1964), p. 59. For more on Mulla Nasrudin, see also Warren S. Walker and Ahmad E. Uysal, *Tales Alive in Turkey* (Cambridge, Mass.: Harvard University Press, 1966).

following tale illustrates the tendency of people to think in habitual patterns which may prevent them from grasping new points of view.

> *Nasrudin used to take his donkey across a frontier every day, with the panniers loaded with straw. Since he admitted to being a smuggler when he trudged home every night, the frontier guards searched him again and again. They searched his person, sifted the straw, steeped it in water, even burned it from time to time. Meanwhile, he was becoming visibly more and more prosperous.*
>
> *Then he retired and went to live in another country. Here one of the customs officers met him years later.*
>
> *"You can tell me now, Nasrudin," he said, "Whatever was it that you were smuggling, when we could never catch you at?"*
>
> *"Donkeys," said Nasrudin.*[13]

SUFI ORDERS AND BROTHERHOODS

Sufism, which began in isolated individual experiences, had, by the thirteenth century, developed into a mass movement. It acquired an institutional structure within which full-time teachers instructed the lay or uninitiated. The different teachings of important mystics formed the basis for orders or brotherhoods that still exist today and that are usually known by the name of their founders. Each order revolves around a specific *tariqa*, or path laid down by a famous mystic, who is often regarded as a saint because he is considered an intermediary between human beings and God. In North Africa such individuals, be they alive or dead, are called *marabouts*. Allegiance to one or another of these orders may cut across different classes and ethnic backgrounds. Women may also participate, although they usually have their own separate groups. Some orders may be associated with segments of the ruling establishment, as, for example, the military during the early Ottoman period. Others may be limited in their membership to marginal groups in the society.

The sheikhs, or masters, meet regularly with their followers in lodges *(zawiyas)* to study and to perform the spiritual exercises. Those who devote themselves full time to the order may themselves become teachers and establish their own *zawiyas* within a loose association of the order. In this way, the orders come to link nomadic camp to village, and village to city. Dispersed as they are, the orders constitute a vital network joining different regions and populations, even cutting across na-

[13] Ibid.

tional boundaries.[14] Today, for example, the Qaddiriya order of Baghdad has lodges spreading from India to Senegal, including some in countries such as Turkey where they function surreptitiously. In the past and in some areas today, *zawiyas* served as hostels for pilgrims and travelers, as schools for children, and as community centers. *Zawiyas* that house the tombs of local saints are considered sanctuaries, or sacred areas where oaths may be taken and where fugitives seek refuge.

POPULAR BELIEFS: *BARAKA* AND "SAINTS"

The success of Sufi orders was probably due to their capacity to incorporate local beliefs and practices into an overall synthesis of popular Islam. Moreover, the close relationship between the *zawiyas* and the local community and the fact that the teachers in these *zawiyas* were most often of local origin and spoke the native dialect was of considerable help in integrating Islam into local community life. Two pre-Islamic beliefs that found their way into Sufism and that were retained in popular Islam are *baraka* and the cult of saints.

Baraka, which means blessing or grace, is actually a very complex concept that may also be translated as holiness, or the quality of divine blessing. Its counterpart is the Christian concept of divine grace and, to a certain extent, the widespread secular concept of good luck. A person who possesses a great deal of *baraka* may be regarded as a saint, or *walī* (*marabout* in North Africa). Both men and women may possess *baraka,* and this charismatic, presumedly wonder-working power may be either inherited or acquired.[15]

All descendants of the Prophet Muhammad or those of his immediate family are believed to have inherited some of his *baraka,* some more so than others. But *baraka* may also be acquired by individuals favored by God. These individuals become living saints who in their lives exemplify extreme piety and divine grace; they demonstrate their *baraka* through their abilities to bring good fortune, heal the sick, and make miracles. Most founders and many leaders of Sufi orders are so considered. Saints' cults may also develop around individuals whose only distinction is to die in battle or under unusual circumstances, with their *baraka* becoming apparent post-mortem. Unlike Christian saints, Moslem saints are usually recognized in their lifetime, and their designation does not derive from a formal declaration or cannonization by any

[14] For two case studies of orders—one historical, the other anthropological—see John Kingsley Birge, *The Bektashi Order of Dervishes* (London: Luzac, 1937), and E. E. Evans Pritchard, *The Sanusi of Cyrenaica* (Oxford: Clarendon Press, 1949).

[15] For a discussion of Moslem saints and *baraka,* see Ernest Gellner, *Saints of the Atlas* (Chicago: University of Chicago Press, 1969).

Rural Turkish women and children visiting a cemetery on a feast day.

religious institution or church. It derives from communal recognition, reflecting a consensus among followers and members of the community.

Tombs of saints are found throughout the Middle East, both in the cities and in the countryside. Usually constructed of dressed stone, these tombs frequently rise from fields alongside major roads to signal a halting place for passersby. Smaller whitewashed and green-domed structures may be the focus for regular visits by members of nearby villages, while larger, sometimes monumental, edifices attract visitors from all over. These tomb-shrines are invariably differentiated from mosques that may adjoin the particularly important ones. Visitors seek out the tomb of the saint, circumambulate it, pray and meditate, and press bits of wax, cloth, or small pebbles on the walls of the chamber to signify special pleas to the saint. Many tombs are found near sacred springs or groves, sometimes enclosed by low walls made all the more dramatic for the lack of similar greenery in the often deforested landscape. The tombs of founders of major Sufi orders lie at the center of great shrine complexes.

Specialized powers may be attributed to different saints; these range from the ability to cure infertility in women, to treat children's diseases, to ensure success in love, or to cure insanity. More women than men seek out the *baraka* of the shrines. As they are barred from participating in formal mosque prayers, women find special meaning in their relationship to the saints.[16] On a fine day groups of women and their children often pass the time in friendly socializing at the local shrines. Entire families make seasonal outings to favorite shrines; they take along their food, which they may share with other visitors or with the beggars who are often found in the vicinity of major shrines.

[16] Fatima Mernissi, "Women, Saints, and Sanctuaries," in *Women and National Development: The Complexities of Change*, edited by the Wellesley Editorial Committee (Chicago and London: University of Chicago Press, 1977). pp. 101–112.

The *baraka* associated with saints has its negative counterpart in the concept of "ill-purpose"—the evil eye or *'ayn*. This force is believed to bring sickness and misfortune and may be thought of as a form of witchcraft.[17] Just as some people have inherent *baraka*, others have in them the power of the *'ayn*. These people, it is believed, can cause bad luck simply by their glance. A person may deliberately direct the evil eye against enemies or their property, but it may also occur unwittingly through unconscious envy. To guard against the evil eye, various amulets and charms may be worn, especially by children, who are thought to be most vulnerable. Valuable animals, such as prized cattle, camels, horses, and rams, are usually protected with blue beads. A common charm worn by women is a hand made of gold, silver, or some metal filigree. Verses from the Quran written on paper and worn on the body in lockets are believed to be especially potent—as are blue beads, amber, cowrie shells, and iron. Phrases such as *bismallah* (in the name of God) and *mashallah* (as God wills) adorn buses and walls and are continually on the lips of the people. Any undertaking, whether it is the beginning of a meal, a journey, or any other task, may invoke *bismallah*, whereas *mashallah* prefaces any praise directed at children or other family members, particularly by strangers. In this case, God's name is invoked to neutralize the likelihood of unconscious envy, and the casting of the evil eye on the individual praised.

SUFISM AND MASS MOVEMENTS

Sufism has an institutional expression that remains important in the communal life of the Middle East. Sufi orders or Sufi-inspired mass movements take many forms and perform different functions. Some are devoted to curing the ill; others emphasize the individual mystical experience and have little public role; still others mobilize the masses into political action and, as such, seem far removed from their Sufi roots. The objectives and functions of the hundreds of orders are so diverse that some scholars question the value of lumping all of them under a single label. We cannot detail here the range of variation among Sufi-inspired movements, but we have selected three cases to illustrate the vast scope of this phenomenon.

The Wahabis of Saudi Arabia are an order at greatest remove from our notions of mystic experience, although mystical in their origin. After a turbulent history of violent opposition to instituted authority,

[17] Brian Spooner, "The Evil Eye in the Middle East," *Witchcraft, Confessions and Accusations*, ASA Monograph, no. 9 (Tavistock Publications, 1970), pp. 311–19.

Wahabism, a puritanical reformist movement, has today become the state religion of Saudi Arabia. Our second example, that of the Hamidiya-Shadhiliya order of Egypt, stands in sharp contrast. Eschewing any militant function or political stand, it is a small, primarily urban-based, tightly organized order whose primary function is that of a fraternal religious club and, on occasion, a place of refuge and material support for its 16,000 or so members. The third case, that of the Shabak of northern Iraq, is again different, and illustrates the encystment of a *tariqa* whereby the order becomes closed to outsiders and becomes the exclusive domain of one small population. It is, in fact, their ethnic insignia.

It must be added at this juncture that one state in the Middle East, Turkey, had once banned all religious orders; through an edict of Ataturk, all religious brotherhoods were declared illegal and were abolished in an effort to promote secularization. Today there is an effective retreat from this extreme position in Turkey, and Sufi orders that had persisted underground since the 1920s are now more open in their activities.

THE WAHABIS OF SAUDI ARABIA

Wahabism had its origin in a special *tariqa* proclaimed by its founder, Mohammad ibn 'abd al-Wahab (1703–1787). A student and exponent of Sufism, 'abd al-Wahab left Arabia to study and teach in Iraq and Iran. After extensive travels, he returned to his homeland and began to teach his own *tariqa,* a highly puritanical interpretation that condemned all innovation in belief and ritual subsequent to Muhammad. He specifically outlawed the veneration of saints and saints' tombs.

Expelled from his native region, 'abd al-Wahab sought refuge with a tribal chief, Muhammad ibn Saud. Ibn Saud espoused the new doctrine and undertook its propagation. The movement grew in power as more and more tribes joined; some came peacefully, others were conquered in battle. By the beginning of the nineteenth century, the Wahabis were so strong that they attacked Karbala in Iraq, Mecca, and even Damascus. Alarmed, the Ottoman government sent a special expedition which succeeded in defeating them and forcing them to retreat back to their original area in eastern Arabia.

The movement then entered a period of retrenchment and general decline until 1901, when Abdel 'Aziz inb Saud succeeded in leading a small Bedouin force and captured the oasis of Riyadh, the present-day capital of Saudi Arabia. From his base in Riyadh, Abdel 'Aziz proceeded to enlarge and consolidate his domain, and by 1915 he had become the master of most of Arabia.[18] He then declared himself the first king of the newly formed kingdom of Saudi Arabia, named after his family, and strict fundamentalist Wahabism became the only Islamic doctrine to be tolerated within its borders.

The great wealth derived from oil after World War II and especially in recent decades, together with the massive influx of non-Saudi Arabs and other foreigners, has created social and political contradictions on a scale hard to comprehend. Wahabism, which espouses a nonmaterialistic, almost ascetic way of life, is increasingly hard to reconcile with the vast wealth which has today transformed the lives of all Saudis and has led to the unprecedented riches in the hands of the royal family. The extensive Saudi family retains virtually total control over all sources of political, military, and bureaucratic power. Paradoxically, their claim to legitimacy rests on their espousal and propagation of the puritanical Wahabi doctrine. This legitimacy is increasingly suspect in the eyes of many. The seizure of the Great Mosque in Mecca, the holiest in all Islam, by a group of Moslem zealots in 1979 bears testimony to the growing gulf between ideology and practice in Saudi Arabia.

THE SHADHILIYA ORDER OF EGYPT

The example of the Shadhiliya is best viewed in the general context of orders in Egyptian society. The following discussion is based on a book by Michael Gilsenan.[19] Gilsenan examines the history and evolution of religious orders in Egypt, and specifically one urban *tariqa,* the Hamidiya-Shadhiliya. Religious orders have always been an integral part of urban society, and many are considered to be well within the mainstream of Sunni Islam and are tolerated by the *'ulama.* Such was the importance of these orders that their leaders came to wield great influence in the society in their dual role as educators of the youth and intermediaries between the masses of the population and the rulers, who were most often non-Egyptians: Turkish, Albanians, or Circassians. Until the nineteenth century, the religious orders were very much a part of Egyptian common life, woven as they were into its social fabric.

The profound changes that transformed Egyptian society during the nineteenth century affected the Sufi orders as well. By the twentieth century, the majority of them were moribund and marginal. Gilsenan at-

[18] For a readable account of the formation of the Saudi Kingdom, see H. Philby, *Arabia of the Wahabis* (London: 1928).

[19] Michael Gilsenan, *Saint and Sufi in Modern Egypt: An Essay in the Sociology of Religion* (Oxford: Clarendon Press, 1973).

tributes the decline of the orders in Egypt to their loss of traditional functions and their concomitant inability to respond constructively to challenges of the changing social system. Rapid urbanization, the shift from subsistence farming to cash crops, and the increased dependency on world markets led to the emergence of a landless peasantry and the creation of a wage-earning class. As the state took over the function of education and trade unions slowly replaced the old craft guilds, the leaders of the Sufi orders came to lose their base of moral influence and power. They no longer played a key role in education, and their previous ties to the craft guilds ceased to be important. Their power was further undermined by their loss of revenue under the land-reform laws which confiscated much of their property.

The Hamidiya-Shadhiliya order, founded by Salama Musa (1867–1939), seems to have been an exception to the general decline. His order managed to grow in membership and hold its own against the competing Moslem Brotherhood, which is an explicitly political activist movement that seeks to establish an Islamic state. Gilsenan attributes its success to several factors. One is that its founder, Salama Musa, was himself a member of the new bureaucratic order while at the same time learned in the Islamic tradition. In establishing his order, he drew upon his administrative experience to create an efficient organization in which a cadre of trained and highly disciplined followers controlled each lodge. A charismatic figure and a reputed performer of miracles, Musa could simultaneously appeal to the clerical workers in the modern sector, to wage laborers, and to peasants.

In the difficult environment of urban Egypt, the order provided "elements of mutual support and benefit, of psychological and material security in a fraternal circle built on cooperation and equality."[20] With its strong emphasis on mutual help and personal discipline, the order offered the individual a sense of identity and security within the changing circumstances of modern Egyptian society. A well-organized network of twenty-five or so lodges suited the needs of workers and lower-level clerical employees who often found themselves working away from their relatives and community in a confusing and impersonal setting.

THE SHABAK OF IRAQ

Our third case exemplifies yet another development of Sufism. The Shabak of the plains of northern Iraq are perhaps the last heirs of the original adherents of the Safawi order, a Sufi *tariqa* first established in

[20] Ibid., p. 206.

Shabak women in a courtyard.

the fourteenth century in Azerbijan. Northern Iraq is an area that exhibits marked environmental and cultural heterogeneity. Its strategic location in the border area between the Ottoman and Persian empires and its mountainous terrain have made it a classical refuge area that sheltered diverse groups escaping religious and political persecution. It remains even today a microcosm of the Middle Eastern ethnic mosaic, which we examine in some detail in later chapters.

Many of the different Moslem groups that inhabit the area—be they Turkish, Arabic, or Kurdish-speaking—belong to the Naqshabandi order, which is generally within the mainstream of Sunni Islam and which is found throughout Turkey, Syria, and the rest of Iraq as well. Other local populations, however, identify with orders considered too esoteric to be classified with either Sunnism or moderate Shi'ism. They are collectively referred to as *ghulat*, or extremists. For example, some settled Turkmen communities belong to the Kaka'i order, others, to the so-called Sarliyya order. The Kurdish-speaking Shabak belong to the Safawi order, although it is difficult to identify it today with the original order of that same name.

Historically caught between the two superpowers in the area, the Sunni Ottomans and the Shi'a Persians, and seeking to maintain their distinct identity and political neutrality, these small, weak peasant populations regrouped around extremist Sufi orders. We might suppose that the religious orders provided an organizational framework and a rallying focus for these people. From the vantage point of the mainstream of Sunni and Shi'a society, membership underscored the marginal status of these communities, marking them off as almost a separate caste.[21]

[21] Amal Rassam (Vinogradov), "Ethnicity, Cultural Discontinuity and Power Brokers in Northern Iraq: The Case of the Shabak," *American Ethnologist*, 1, no. 1(February 1974), 207–18.

Leaders of the Shabak, known as *pirs,* played political and educational roles quite different from those of their counterparts in Egypt. The Shabak, numbering around 20,000, were until the mid-1950s a landless, sharecropping, illiterate people working the fields of urban landlords, who were Sunni Arabs. The Shabak had in effect two classes of leaders: their *pirs* served their spiritual needs and functioned as leaders in internal communal matters; their ethnically distinct and powerful landlords served as general patrons and represented them vis-a-vis the government.

In common with other Moslem-derived secret sects, the Shabak have developed their own interpretations and ceremonials. Their distinctive features include private and public confessions, tolerance of alcoholic beverages, and a general laxity in observance of Moslem ritual obligations. The primary means of social integration within the community is that of the relationship between adult males and their spiritual guides, the *pirs.* The relationship to a *pir* tends to be inherited, so that lay families are associated with their *pir* families over generations. The *pirs* themselves are grouped into several levels all under the leadership of the supreme head of the order, the *Baba.* The religious calendar is crowded with private and public ceremonies presided over by *pirs.* These occasions bring together kinspeople and neighbors who participate in public sacrifice and the sharing of communal meals, although rarely are there true feasts due to the general poverty of the Shabak.

The Shabak represent an extreme instance of a closed or encysted community in the Middle East, although they have parallels among other impoverished rural groups in Turkey, Syria, and elsewhere. Their secretive beliefs, regarded as heretical by outsiders, their insistence on marrying within the community, and their rural isolation define the boundaries separating them from the rest of the society. Their elaborate socioreligious system, which has helped to maintain the group over time, is also the means of their exploitation. As long as they were viewed as Shabak, they were largely denied access to routes of mobility in the larger society, which reinforced their dependency on their landlord-patrons. Only with the recent land reforms and the direct intervention of the revolutionary Ba'athist government of Iraq have things begun to change for the Shabak. Some now own their fields; young men join the army; and still others go to work in the factories nearby.

These three cases indicate some of the divergent ways in which Sufi *tariqas* have developed and their quite distinct functions in different social and political contexts. Of course, Sufism is not limited to its expression in any specific *tariqa* or order, or even in any specific set of beliefs and practices. Rather, at its core it represents a shared attitude concerning the individual and his or her relationship to God.

ISLAM AS CULTURE

In this and the preceding chapter, we have discussed the origins of Islam in western Arabia, its core of belief and distinguishing ritual. We also sketched the history of its transformation into a world religion embracing different peoples and a variety of cultural traditions. We noted its cleavages and the variety of its expressions in the political and social spheres. However, this is not a book on Islamic civilization, let alone on the modern history of the area. What we hope to convey here is a sense of Islam as part of the daily life of the people and something of its role in Middle Eastern society. In short, we are interested in Islam primarily as a living cultural tradition in the Middle East.

All too often today the term Islam is invoked to "explain" a whole range of phenomena. These include political instability, oppression of women, economic underdevelopment, national xenophobia, and a host of psychological attitudes such as fatalism, rigid conservatism, and dependency. In fact, such a simplistic perspective takes us back to an earlier period in history, a time when the Christians and Moslems vied with each other to control the Mediterranean and viewed each other in essentially religious terms.

Few, if any, would invoke Christianity to explain all the features of Western society; Islam, likewise cannot be considered as determining all the features of society in the Middle East. However, as with Christianity and Judaism, Islam is a shared set of symbols and meanings which people use to identify themselves, to impart meaning to their lives, and to express certain aspirations. Islam remains the single most important source for the ethos that distinguishes the area and imparts to its bewildering complexity and variation a measure of unity and cultural uniformity.

Having said this, we have to say that Islamic culture in the Middle East, however defined, is shaped by its own regionally varied history as well as by its present realities. At this level, many have even found it useful to speak not of "Islam" but rather of "Islams," in order to underscore the existing variations in time and place. Just as Islamic culture in Iran is different from Islamic culture in Egypt, Islam within Iran is expressed differently and it plays different roles among the different segments of that society, be they defined in terms of region, ethnic groups, or economic class.[22]

As we suggested earlier, scholars have found it useful to approach Islam (like any other world religion) in terms of analytical distinctions

[22] Abdul Hamid el-Zein, "Beyond Ideology and Theology: The Search for the Anthropology of Islam," *Annual Review of Anthropology*, 6 (1977), 227–54.

between Great Tradition/Small Tradition, and Formal/Folk (or Popular) Islam. Clearly, the presence of urban-based traditions of scholarship, artistry, and learning has created within Middle Eastern society distinctive approaches to religion as a system of meaning and symbols. In many respects this forms a cultural barrier that separates the literati from the masses, the administrators from the governed, the landlord from the sharecropper. What the anthropological perspective on Islam as culture might best offer is not further analytic distinctions which, in effect, tend to establish normative Islam in opposition to Islam in practice. Rather, it should encourage us to see Islam as being simply what people who profess it believe and do. In this sense, Islam is neither moribund nor a relic of the past any more than are the people themselves. Islam can be as much a form of revolution as it is of conservatism, a power for justice as well as a tool for oppression. It can liberate as well as constrain. In Iran a despotic regime was overthrown by the unity forged by the Shi'a 'Ulama among disparate segments of society through shared Islamic symbols. At the same time, the future of the Iranian Revolution and the territorial integrity of the country are threatened by conflicting interests and even conflicting views as to what constitutes "the real Islam."

CHAPTER FOUR
COMMUNAL IDENTITIES AND ETHNIC GROUPS

It should be clear by now that the contemporary society of the Middle East is the product of a long and varied historical process, one of which the people themselves are acutely conscious. A strong sense of historical awareness often serves to validate the present as it provides a focus for unity and solidarity. Paradoxically, this historical awareness serves to differentiate one group from another in the area and therefore contributes to cultural diversity and disunity. This diversity, visually expressed through distinctive styles of dress, ritual, and public behavior, must be properly appreciated if we are to understand Middle Eastern society. We have already seen in our discussion of Islam some of the historical processes that differentiated groups within the Islamic tradition; these differences, or cultural cleavages, themselves parallel, amplify, and even define group boundaries and structure intergroup relations today.

Carleton Coon likens this diversity to a "human mosaic," with the members of each identifiable group emphasizing their common and special identify through some configuration of symbols. These symbols may be material—in the form of dress, dwelling styles, language or dialect; of even greater significance, however, are the underlying patterns of behavior, values, and systems of belief. The recognition and acceptance of ethnic or communal differences have traditionally been a fundamental principle of Middle Eastern social organization. The notion of a human mosaic, however, has its limits. Although it describes contemporary patterns, it offers little insight into the historical processes that underlie the formation of group identities, how these change over time, and, more important, how people use them to gain access to resources and power.

Until the rise of nationalism, most polities comprised aggregates of bounded social groupings, be they tribes or confessional communities. Whether joined for common purpose or held together by threat of force, the distinctive quality of the individual grouping was maintained through the principle of collective responsibility. It is interesting to note that more than a century of Turkish, Arab, and Iranian nationalistic movements have not succeeded in eroding the significance of more narrowly defined ethnic identity for the individual. In this chapter we discuss the broad outlines of ethnicity and the sources of individual and group identities.

Each country and region of the Middle East contains local groupings or populations which are distinct from the society as a whole and which are recognized as such by themselves and others. That is to say, people recognize themselves as belonging to some unique grouping within a larger population. Some of these groupings may be called *ethnic,* following Schermerhorn's definition: "... an ethnic group is a collectivity within a larger society having a real or putative common ancestry, memories of a shared historical past, and a cultural focus on one or more symbolic elements defined as the epitomy of their peoplehood."[1] The elements used to signal the identity of ethnic groups include religious affiliation, language or dialect, tribal membership or shared descent, racial variation, and regional or local customs.

ETHNICITY: A THEORETICAL FRAMEWORK

In considering ethnicity in the Middle East, we should keep in mind a number of points which together make up a framework for the understanding of ethnic group relations. First, ethnic groups are ultimately defined in contrast to other groups. As a consequence, individuals under different circumstances identify themselves in different ways. One Turkman will distinguish himself from another Turkman by membership in a particular tribe or section. For example, he may be con-

[1] R.A. Schermerhorn, *Comparative Ethnic Relations: A Framework for Theory and Research* (Chicago: Chicago Press, Phoenix Edition, 1978) p. 12.

Three generations of men in a family from upper Egypt. (Photo courtesy of Diana de Treville)

scious of his Goklan identity when dealing with a member of a nearby Yomut village. In town, in transactions with Iranian or Azeri shopkeepers, this distinction is immaterial, and he is simply a Turkman. Should a Christian ask the Turkman who he is, his response would most likely be to term himself a Moslem. In other words, ethnicity here is simply the basic "me" versus "them" distinction, and the definition of each and the boundaries vary with the context.

Individuals may also consciously or unconsciously choose among a number of alternative sources of social identity in efforts to manipulate social or economic situations. One may adopt the language, symbols, and codes of a special grouping to which one's ties are quite remote or even nonexistent. In this sense, ethnic identity may be considered as a personal strategy, a means to accomplish a desired objective. For example, in Iran prior to the 1979 Revolution, educated and well-to-do members of different non-Persian-speaking groups such as the Qashqa'i or the Azeri were assimilated into the national elite using Persian upper-class speech mannerisms and social codes. They nonetheless usually maintained their original cultural identification when with their own people.

Even individuals who have no desire to assimilate or "pass," as it were, frequently use the codes and symbols of others to facilitate communication or to simply show respect. Of course, there are both psychological and social limits as to how one can use or manipulate ethnic or other forms of group identity. Individuals are socialized into primary groupings in ways that encourage a psychological commitment to their close relatives and to the symbols and values to which they adhere. Rarely do individuals repudiate these primary ties. Moreover, there are practical constraints on the manipulation of sources of identity. One constraint has to do with the willingness of other people to accept one's use of a particular identity. If one too blatantly shifts identity and allegiance, one also risks loosing the long-term support of his or her primary group.

Although ethnic identity is ultimately an individual strategy, its main social significance emerges to the extent that it serves as the basis for political or economic organization. Throughout the Middle East, there is a strong tendency to find occupational specialization associated with particular ethnic groups. For example, the Jewish community in Isfahan was important in fine metalwork and in trading of gold and silver: Kurds in Istanbul and Ankara have a near-monopoly as porters in the bazaars; most hotels and restaurants in Iraq are run and staffed by Assyrian Christians; most of the long-distance truck drivers of Iran and automotive mechanics are Azeri Turks; and most of the cooks in Egypt are (or were) Nubians.

Although it is not possible to identify particular tasks or occupations exclusively with particular groups, there are some general associations. For example, gypsies are closely identified with tasks thought by Moslems to be polluting or degrading, such as dealing in animal hides and public entertainment. Christians and Jews have long been associated with forms of commerce and business which, for either religious or other cultural reasons, were felt unsuitable for Moslems. Money lending or money changing and import-export activities which relied on such transactions were dominated by non-Moslems until very recently. The role of non-Moslems in such trade, of course, was furthered by their ability to use their contacts with coreligionists outside the area.

What is almost universal is for each region or community to have its locally unique patterns of division of labor along ethnic lines. There are organizational advantages in having skills and trades passed from father to son, and there are advantages in closely related individuals, following the same craft or line of trade. Quite apart from facilitating training, in the absence of national banking institutions, close compatriots are often sources of credit or capital given on the basis of personal trust and reputation. Communication is also easier among kinspeople, which probably facilitates the local prominence of one or another group in a particular endeavor. Because lines of patronage and mutual support often reflect primary group ties, it is not surprising that as new jobs or employment opportunities arise, they may be filled by people sharing a social or ethnic identity.

Fredrik Barth, an anthropologist who has worked in Iraq, Iran, Pakistan, and Oman, has drawn attention to an important aspect of ethnicity. What is most important, he suggests, is how ethnic-group membership structures access to resources and intergroup relations. In other words, Barth shifts attention from the cultural content of ethnicity—that is, the symbols and codes which define it—to how it facilitates or impedes the access of people to resources. In Barth's view, ethnic groups can be thought of as occupying unique places in the social landscape, much like the niches occupied by animal species in a particular ecological system.[2] Unlike the mosaic image, which is essentially static, the niche of any group is determined by what they do for a living, their social and political organization, and their relations to other groups in their environment. Further, this approach draws attention to the fact that occupational or productive specialization on the part of an entire group can be a very effective means of utilizing available resources, including labor and acquired skills.

This model emphasizes the complementarity of the roles and functions served by the different groups who interact with one another. For example, the Yörük of southeastern Turkey, about whom we will have more to say later, have traditionally been nomadic pastoralists. They do not, however, own pastures and must acquire grazing rights from local landlords and villages. Thus, their niche is defined as much by the activities of other groups as it is by the needs of their animals and the grasses which sustain them. The Yörük exploit marginal areas and high pastures which local farmers are not equipped to fully utilize. The close lines of communication and mutual support among the Yörük make it difficult for outsiders to effectively compete with them in getting pastures, in organizing sales of animal products, and in moving flocks. Ethnicity, with its emphasis on shared unique social characteristics, thus facilitates access to certain resources, even the defense of them against others. As a result, particular resources or ways of exploiting them may become identified in many regions with particular peoples.

However, we have to keep in mind that complementarity or mutuality of benefits is only one aspect of intergroup relations. Groups frequently establish exploitative relationships with others in which ethnic identity may serve to organize and perpetuate inequality. The Shabak of northern Iraq, for example, were a caste-like grouping of agricultural sharecroppers who depended on politically and socially

[2] Fredrik Barth, "Ecological Relations of Ethnic Groups In Swat, North Pakistan," *American Anthropologist*, 58 (1956), 1079–89. For a broader theoretical framework of ethnicity, see Barth's Introduction to *Ethnic Groups and Boundaries: Social Organization of Cultural Differences*, ed. Fredrik Barth (Boston: Little Brown, 1969).

dominant urban Arab landlords. Differentiated by language and religious practices, the Shabak remained a weak and exploited group. As Shabak, they were systematically denied access to better jobs outside their community and found it difficult to find anyone who would sell them land in the past. The ethnic label of Shabak locally connoted poverty, backwardness, and low status. Should a Shabak family acquire wealth or move into town, it would rapidly try to disassociate itself from the rural community, which at times included changing its language and customary practices.

Even when a high degree of economic mutuality exists among members of different groups interacting together, there may also be considerable mutual antipathy. For example, the Sulubba, gypsylike nomadic peoples of Arabia, specialize in metalwork, music making, and entertainment. They regularly move from one Bedouin camp to another or between villages, plying their trade. Despite close association with their hosts and clients, they are held in low esteem and are socially ostracized. Although at times the very values and attitudes held by members of a particular ethnic group towards others may engender overt hostility and conflict, this should not be overly stressed as a generalization. Mutual accommodation and tolerance are by far the more common basis for communal interaction.

We should reemphasize that ethnicity is an analytic concept used to describe or understand aspects of individual or group identity. While we speak easily of particular ethnic groups and their cultural boundaries, we have to keep in mind that the effective units of social action implied by ethnic labels are ever changing. Whether or not a particular ethnic category of people or identifiable collectivity is meaningfully thought of as a "group" depends on a knowledge of the specific circumstances. For example, it is not useful to regard Arabs in the Middle East as an ethnic group. However, a small subpopulation of Arabic speakers within a Persian-speaking community *may* interact in a way that defines a bounded group in the way that Schermerhorn and Barth suggest.

In the following sections of this chapter, we look at important sources for cultural differentiation which, while independent of ethnicity, may be utilized at any given time to define ethnic-group boundaries. These sources for differentiation constitute the raw materials for ethnic identity and group formation.

RACE

Of all the elements that may be used to define groups or social categories, phenotypic race is the least important in the Middle East, where the vast majority of the people from Egypt to Afghanistan tend to fall within the same racial grouping, often referred to as Mediterranean. Where a markedly differentiated population exists, such as the *'abid* or blacks of Saudi Arabia, the Nubians of Egypt, or the Turkmen of Iran (with pronounced Mongoloid features), the recognition of phenotypic differences is locally associated with an ethnic identity. Even though in much of the area light skin is considered a mark of beauty and higher status, there is no prevailing ideology of race based on color. Until quite recently slavery practiced throughout the Islamic world was not exclusively associated with Africans or any other people. In the Arabian Peninsula, as might be expected by virtue of geography, most slaves were East African in origin, and their descendants still form fairly distinct groupings within Peninsular Arabian society. The Ottomans recruited slaves from both Eastern Europe and the Caucasus. In general, the descendants of these slaves today do not form either racially or ethnically distinct groups. This is because many of the "slaves" were employed in high-level administrative positions and in the military. In fact, they were not slaves at all in the sense of chattel, but rather were part of the sultan's entourage and administrative cadre. Once converted to Islam, such individuals served the government throughout the empire, married, and accumulated property. Slavery in this case meant little more than "servant of the sultan."

Outside a few towns in southern Arabia, slavery in the Middle East has never been a primary means of organizing menial labor. Perhaps as a consequence, the association of class and race or ethnicity and race is not well developed in the area.[3]

LANGUAGE

As we have noted, under certain circumstances linguistic differences can become ethnic markers. But more important, language serves to establish boundaries on a much larger scale. The three major language groupings in the modern Middle East are the Semitic, Indo-European, and Altaic or Turkic linguistic families. These are broad classifications, and each encompasses a number of major languages and numerous distinctive dialects. Arabic and Hebrew are Semitic languages. Whereas Hebrew is spoken only in Israel, Arabic is the national language of Egypt, Lebanon, Syria, Jordan, Iraq, Saudi Arabia, and the smaller states of the Arbian Peninsula. The Indo-European language family is regionally represented by the many dialects of modern Persian, Kurdish, Luri,

[3] Bernard Lewis, *Race and Color in Islam* (New York: Harper Row, 1971). Although somewhat dated, Albert Hourani's book, *Minorities in the Arab World* (London: Oxford University Press, 1947) remains a valuable source on the subject.

Baluchi, and smaller groups of Greek and Armenian speakers. The major Turkic languages and dialects are western or standard Turkish of Anatolia, Azeri of Iran, Turkmen, and the languages of other smaller groups of Central Asian origin, such as Tatar and Kazak.

Language in and of itself usually establishes only the outermost parameters to group membership, although dialect differences may quite precisely identify a person as to region or even tribe. In Baghdad, for example, Moslems, Jews, and Christians all speak Iraqi Arabic, a dialect distinct from colloquial Egyptian or Syrian. However, the Moslems, Jews, and Christians of Baghdad can be distinguished from one another on the basis of distinctive speech mannerisms, syntax, and grammar.

Spoken Arabic represents a number of distinct speech communities that vary regionally; those of Egypt, Syria, Iraq, and Saudi Arabia are quite distinct. Even within the countries there exist regional variation, as, for example, between lower and upper Egypt. In the extreme southern part of the Arabian Peninsula, a small number of communities speak a highly variant dialect of Arabic identified as Southern Arabian. Nonetheless, the major dialects of Arabic are usually mutually comprehensible and are all written in one form. The rapid development of mass communication and the extention of public education are facilitating the breakdown of dialectical barriers and encouraging the spread of a common standardized Arabic used in publication everywhere.

Persian, an Indo-European language, is the national language of Iran and encompasses many closely related dialects spoken as a first language by about 23 million people in Iran and by about 621,000 people in neighboring Afghanistan. Persian is written in Arabic characters and has a substantial Arabic vocabulary. The infusion of Arabic is due in part to the early politico-religious domination of Iran by the Arab Moslems, and at least in equal part to the use of Arabic by medieval Persian scholars and men of letters. As with Arabic, dialectical variations in Persian serve to differentiate class and regional affiliations.

The third national language in the Middle East is Turkish. Of the nearly 90 million Turkic-speaking peoples in the world, some 35 million live in Turkey proper where dialectical differences are relatively minor when compared with other primary language families. Urban-rural differences in speech frequently overshadow regional differences, although certain interregional linguistic differences can be easily distinguished as, for example, the Black Sea Coast from the Mediterranean. Until the reforms of the Ataturk period, Turkish was written in the Arabic script; since 1928 Turkish has been written in the Roman alphabet.

So far we have talked about Arabic, Hebrew, Persian, and Turkish, which are national languages. In every state, however, there are important communities that speak other languages, as well as a substantial amount of bilingualism. This gives a political dimension to language and constitutes the level at which language is most salient in defining ethnic boundaries. In Turkey, for example, as many as 2.5 million people are native speakers of Kurdish, and many other small groups of people speak Armenian, Greek, Ladino (a Spanish dialect spoken by Sephardic Jews), Tatar, Circassian, and Bulgarian. Perhaps as many as 100,000 people in Turkey speak Arabic. However, only standard Turkish can be used in schools, courts, and the media. Kurdish newspapers, books, and records are illegal and, at one time, even the mention of the term Kurd or Kurdish was discouraged. However, in neighboring Iraq, where the Kurds form an even larger minority within the population, Kurds have their own radio station and press, and Kurdish is taught as a second

Iranian women from Isfahan.

language in districts where Kurds predominate. This has not been the case in Iran, where some 2 million Kurds live along the northwestern border. Although an Indo-European language, Kurdish is grammatically and lexically distinct from Persian. Public education is conducted exclusively in Persian, and the use of Kurdish in public media is restricted.

Like Kurdish speakers, the Baluch are also divided among a number of nation-states: Iran, Pakistan, and Afghanistan. The Baluch, who number around 3 million, speak highly localized variants of Baluchi, an Indo-European language. Located in one of the most arid mountain zones of southwest Asia, most of the Baluch eke out a living as nomadic pastoralists, coastal fishermen, and farmers.

The political implications of the presence of these large, linguistically differentiated minorities, like the Kurds, Baluch, Turkmen, or even the encapsulated Arabic-speaking communities in Turkey and Iran, are touched upon later in this chapter. What is important to consider here is that in every country there is a strong nationalistic movement underwriting the use of particular languages or dialects to promote unity in the face of considerable and deeply rooted cultural diversity. For members of local populations, like those we have noted, this often presents a major dilemma. To participate fully in the national economy, to educate their children, and to partake fully in the emerging national culture, they have to acquire a second language and dissociate themselves to some extent from their primary communities.

Although the educated elite in most countries of the Middle East is almost always bilingual in a European language, great emphasis is placed on the promotion of a national tongue among the masses. Bilingualism in French or English serves as a sign of education and status, but with the possible exception of Lebanon, neither French nor English is utilized in public transactions. This is in marked contrast to North Africa, where French is still both the private language of the middle and upper classes as well as the language of trade and government. Even in Algeria, with its long record of anti-French struggle, the elite is essentially French-speaking.

RELIGION

Although language and local dialects are significant in the differentiation of people and may delineate distinctive communities, religion is the most important single source of personal and group identity—and, by extension, social cleavage. The perceived rights and obligations of one person to another are strongly tempered by whether or not the parties involved are coreligionists. The assumption that, in the final analysis, an individual will turn to and favor others of his or her faith is so pervasive as to constitute a basic principle of social interaction. Religion in its many sectarian expressions sets some of the most important limits to interpersonal behavior. Of all injunctions regarding marriage expressed by the religions of the Middle East, the one fundamental to all is that marriage be restricted to coreligionists.

This is not to say that religious ties or bonds inevitably supersede all others. In fact, class differences, tribal and ethnic divisions, and the like often take precedence over any claims of religion as people organize themselves in groups for common action. Religion is more a determinant of maximal boundaries or inclusivity, less commonly the basis for local organization. For example, as is sometimes the case in southwest Turkey, a village may be comprised of both Sunni and Shi'i residents. This distinction will almost inevitably be reflected in residential segregation. However, within each residential quarter. groups organizing for political action or other purposes are most likely to utilize more exclusive criteria for membership. Recruitment for political action is more apt to be along lines of common descent and tribal affiliation. One has only to look at the persistent conflicts in Iran between the Shi'a Azeri and the Shi'a Persians, or between the Sunni Kurds and Sunni Arabs of Iraq for examples of ongoing intrasectarian conflict. Moreover, ties of close friendship, contractual partnerships, and political alliances everywhere join people across sectarian boundaries. Residence within towns and cities, while always expressing some aggregation along sectarian lines, is rarely homogeneous. In virtually every big city, for example, Christians, Jews, and Moslems live in close juxtaposition, and even share apartment buildings and compounds.

Although religion cannot be evoked to explain all or even most patterns of social interaction, sectarianism continues, nonetheless, to be a factor with important political and social consequences for every country. Nationalistic movements both within individual countries and those, like Pan-Arab nationalism, which transcend state frontiers have consistently found it difficult to reconcile sectarianism with their more encompassing national political objectives.

NON-MOSLEM CONFESSIONAL COMMUNITIES

The tenacity of sectarianism in the political arena can only be understood when considered within its historical context, because the nature of religious distinctions in the Middle East and the development of specific confessional communities are rooted in the formative period of Islam. Furthermore, all successor states and governments have, one way or another, perpetuated the idea of the confessional community as part of the structure of society.

From its inception, Islam as a faith was explicitly the basis for political action; even within the lifetime of its founder it became the vehicle for the formulation and organization of a new polity. The boundaries of the political community were not simply territorial, but coincided with those of the religious one, or *'Umma*. Both the legitimacy of the ruler and the rights and responsibilities of the members derived from their common membership in the religious community; it followed that to be a full citizen was also to be a Moslem. Although over the centuries there have been many Moslem states, and although many Moslem groups have been encapsulated within non-Moslem polities, recognition of common membership in one *'Umma*, or Moslem community, remains a potent political and ideological force.

What then of the non-Moslems who were incorporated into the early Islamic empires and their successor states? The Prophet Muhammad regarded the Christians and Jews as "The People of the Book," the recipients of a valid but incomplete, and hence imperfect, revelation. Members of both communities were allowed to practice their religion and keep their institutions and property, but on the condition that they pay a special poll tax. They were not allowed to serve in the army or assume direct authority over Moslems. As a consequence, Christians and Jews assumed a status of tolerated clients of the Moslem community—clients who suffered certain sociopolitical liabilities in exchange for protection and the right to retain their distinctive communal identities.

While the significant distinction is between Moslem and non-Moslem, the underlying principle is more widely employed. Religious identity, even as narrowly defined by the sects that rapidly arose within Islam itself, assumes political significance. The systematic merging of social and religious identity persists and has become even further institutionalized with time, as we see, for instance, in the Coptic-Moslem social cleavage in Egypt.

Within the Ottoman Empire, the immediate political predecessor to many of the modern states with which we are concerned here, this system was particularly developed, as it formed a principle of the organization of the Empire. In May, 1453, the Ottoman armies led by Sultan Mehmet II captured the city of Constantinople from the Byzantines. Even though the Ottomans already ruled vast lands containing Christian subjects in Europe and Anatolia, it was only after the fall of Constantinople that they institutionalized a system of governance that came to be called the *millet*. The Sultan sought the support of the Christian religious leaders of the city and assured the Greek Orthodox clergy that it could retain civil as well as religious authority over Orthodox Christians in the Empire.[4] This practice of delegating considerable authority to community and religious leaders is the basis of the *millet* system, which was later extended to the Armenians, Jews, and others. Quite apart from its being an effective way of ruling non-Moslems, it had the effect of reinforcing the political and social significance of sectarian identity. Marriage, divorce, and other aspects of personal status were all regulated through the *millet*. The *millet* was also responsible for resolving many of its own internal conflicts and for paying some taxes collectively. By the nineteenth century, seventeen different *millets* were recognized by the Ottoman government.

We can see that religious or communal cleavages are basic to the structure of Middle Eastern political life and are not simply anomalies or the result of imperfect assimilation. The various Christian sects and the Jews have, over centuries, accommodated themselves to Moslem rule, just as the Moslem majority has recognized their right to persist in a separate communal order. Within Islam itself, the various sects often maintained a separate communal order resembling that of the non-Moslem minorities.

Only with the advent of nationalistic ideologies in the nineteenth and twentieth centuries did this communal and sectarian compartmentalization come to be questioned and tested. The issues of what constitutes nationality and full citizenship and how to accommodate sectarian differences still have to be resolved in most of the states of the area. Interestingly, this is as true of Israel as it is of the other states. To appreciate this dimension of Middle Eastern society and politics, we will describe briefly some of the major distinctive communities in the region. We begin our discussion with the Jews, who form some of the oldest continuing ethno-religious communities in the area.

Until the establishment of the state of Israel in 1948, a number of fairly large Jewish communities were found throughout the Middle East. On the whole, these tended to be urban with a few rural exceptions, such as the Jews who lived in villages in northern Iraq. The latter spoke a Hebrew dialect, *targum,* and were the clients of powerful Kurdish chiefs in the area. Ranging from very wealthy urban bankers and merchants with international connections to destitute, small shopkeepers, the members of the Jewish communities of the Middle East reflected in their lifestyles the different cultural traditions of the areas in which they lived. While displaying great internal diversity both in language and culture, most Jewish people residing in the Middle East prior to World

[4] Stanford J. Shaw, *History of the Ottoman Empire and Modern Turkey* (Cambridge, Eng.: Cambridge University Press, 1976).

War I were either dispersed remnants of ancient Jewish communities or were members of mostly Sephardic populations that, fleeing European oppression, sought refuge in Ottoman and other Moslem lands. In places the Sephardic Jews preserved their Spanish heritage through the use of Ladino, a Spanish dialect. The Ladino-speaking Sephardim were mostly urban and were concentrated in Istanbul and Izmir in Turkey.

In 1950 there were about 130,000 Jews living in Iraq and some 30,000 in Syria, the Syrian Jews being largely Sephardic.[5] The Jews of Iraq were divided into distinct classes, with a small wealthy group dominant in commerce and banking. There was also a large professional class which, prior to Iraqi independence, occupied important positions in government service and commerce. The great majority of very poor people engaged in petty trading and handicrafts, with the exception of the so-called Kurdish Jews, or *targum*, who were scattered in agricultural groups.[6]

Cairo, like Baghdad, also had a large indigenous Jewish population and still boasts one of the oldest synagogues in continuous existence in the world. Here again the Jewish community, while distinguished by class differences, resided primarily in one quarter of the city and engaged in commerce, artisanship, and peddling. The 1947 Egyptian census listed 65,639 Jews.[7] The position of the Jews in Egypt, like that of the Jews in Syria and Iraq, became extremely uncomfortable during and after the 1948 War of Independence and the establishment of Israel. Moreover, "many Jews were closely associated with the British, French, Italian, Greek and other foreign communities. When nationalist sentiment became inflamed against Israel and the West, it also rose against the Jewish community, finally bringing about a mass exodus after the British, French and Israeli invasion in 1956." [8]

There have been Jews in Iran since ancient times, and even though their native tongue is Persian, they maintain a strong sense of separate identity fostered by close intermarriage, residential segregation, and a focus on a number of shrines and pilgrimage centers within Iran, notably in Yazd, Isfahan, and Hamadan; a major shrine is the tomb of Daniel in Shush.[9]

[5] Walter Zenner, "Syrian Jews in Three Social Settings," *The Jewish Journal of Sociology*, 10 (1968), 1–22.

[6] Moshe Seltzer, "Minorities in Iraq and Syria," in *Peoples and Cultures of the Middle East*, ed. Ailon Shiloh (New York: Random House, 1969), pp. 10–50.

[7] Gabriel Baer, *Population and Society in the Arab Middle East* (London: Routledge Kegan and Paul, 1964), p. 90.

[8] Don Peretz, *The Middle East Today*, 2nd ed. (Hinsdale, Ill.: Dryden Press, 1971), p. 147.

[9] Herbert Vreeland, "Ethnic Groups and Languages of Iran," in *Peoples and Cultures of the Middle East*, pp. 51–68.

The various Jewish communities of Middle Eastern origins form a distinctive segment of Israeli society and are collectively known as *mizrachim*, or Oriental Jews. They make up about 50 percent of the population of Israel, yet are still poorly represented in the upper echelons of the government, army, and bureaucracy. Some have become identified with particular jobs; many of the Israeli police, for example, are Iraqi in origin. Yemeni Jews brought with them highly prized skills in metalwork and embroidery, and they tend to be employed as skilled artisans. Other Oriental Jews have not done as well, because they lack skilled crafts or education.

The Christian communities in the Middle East have a long and turbulent history.[10] Originally, all Christians in the area belonged to one or another of the indigenous churches that followed the Eastern rites. Outside the Greek Orthodox Church, which was the official church of the Byzantines, the others had their origin in the schismatic "heresies" of the fifth and sixth centuries. Two of the largest Christian communities, the Copts of Egypt and the Assyrian Nestorians, go back to the fifth century and the religious controversies that culminated in the Council of Chalcedon, A.D. 451. The disagreements ostensibly had to do with dogma, specifically that concerned with the nature of Christ. However, these secessionist movements also represented efforts by the local populations of the area to free themselves from the cultural and political domination of the Byzantines. In fact, the oppressiveness of Byzantine rule must have greatly facilitated the early and rapid success of the Moslem armies. Rapid capitulation and massive conversions of Christian rural populations dispersed the Christian communities.

Rural-dwelling Copts are concentrated in Upper Egypt, where they are found in villages little distinguished from those of their Moslem neighbors except for the presence of a small church. Copts follow their own religious calendar, with its periods of distinctive fasting and holidays. Their clergy, who are allowed to marry, wield considerable power in their communities, where they act as leaders. Like the Moslem community at large, Copts prefer arranged marriages among close relatives—for example, the marriage of the son to his father's brother's daughter. Copts speak Egyptian Arabic and are culturally very much like the Moslem Egyptians of the same social class and education.

The Assyrian Christian community of Iraq, Syria, and Turkey, like the Copts, once constituted a *millet* within the Ottoman Empire. Al-

[10] For an overview of the history of Christian communities in the area, see Robert Haddad, *Syrian Christians in Muslim Societies; An Interpretation* (Princeton: Princeton University Press, 1970); see also F. W. Hasluck, Christianity and Islam Under the Sultans, 2 vols. (Oxford: Oxford University Press, 1929). Also see Otto Minardus, *Christian Egypt, Faith and Life* (Cairo: The American University in Cairo Press, 1970).

though today the majority of them are probably to be found in the United States and Canada (with some recent immigrants to Sweden as well), enough remain in these countries to make up distinctive minorities. The largest is that of Iraq. In origin, most of the Assyrians were rural dwellers in the northern part of the country; today they are found throughout the cities and towns of Iraq, with the largest concentration in the capital of Baghdad.

The majority of the Assyrians belong to the Nestorian Church, with a small group divided among the Catholic and Protestant Churches. The Nestorian Church was originally centered in southern Iraq, and its communities were scattered over a large area, some as far as central Asia and even China. Until recently, the Assyrians, who speak an Aramaic dialect, were an agricultural people scattered in villages in the mountains of Hakkiari (on the Iraqi-Turkish border) and in the valleys east of Lake Urmia. Caught in the turbulent politics of the area following World War I, the Assyrian leaders allied themselves with the British, and Assyrians joined the British army in Iraq as special levies. With Britain's help, a group of Assyrians hoped to establish an independent state for the Assyrians in northern Iraq. However, the British shift in policy and their interest in the establishment of an Arab monarchy in Iraq together with the divided Assyrian leadership put a cruel end to their hopes. Not only were the Assyrians frustrated in their efforts at independence, but several hundred of them were massacred in 1933 as they tried to flee Iraq into French-held Syria. Today the Assyrians who remain in the Middle East are scattered throughout Iraq, Syria, Turkey, and Iran.

Besides the Copts and the Assyrians, there are other Christian sects of importance in the Middle East. A group of these are known as the Uniate Churches, of which the Maronites of Lebanon make up the largest community. The name *Uniate* refers to those communities of the Eastern Churches that chose to recognize the authority of the Pope and to adopt the Latin rites. They did, however, retain their own patriarchs and internal autonomy. For example, the Chaldean Christians—a group of about 160,000 found in Iraq, Syria, and Iran—split off from the Nestorian Church in 1750 and, urged by Dominican missionaries, joined the Catholic Church, acknowledging the Pope's authority in matters of dogma. Locally recognized as an enterprising and hard-working people, the Chaldeans dominated the service sector in the cities of Iraq, especially the hotel and restaurant business. Following World War II, large numbers of them, like their Assyrian neighbors, emigrated to the United States and Canada.

It is impossible to discuss the Maronite Christians without reference to Lebanon. The plight of Lebanon today illustrates both the consequences of those historical processes that locate politics in sectarianism, as well as the impact of Western colonialism and international power struggles on local politics.

LEBANON AND THE MARONITES

The Maronite Uniate Church is a national one that is in the main limited to Lebanon and makes up the single largest Middle Eastern Christian community outside of Egypt. Syrian in origin, the Maronites are the followers of Saint John Maroun (d. 410), who lived and preached near Antioch in present-day Turkey. Persecuted by the Byzantines in the fifth century, they sought refuge in the mountains of north Lebanon and later spread to north-central Lebanon, which still remains their stronghold. Ruled directly or indirectly by the Ottomans from 1516 until the establishment of the French Mandate in the 1920s, the Maronites managed to retain a measure of autonomy which varied largely in response to the support they received from European powers. By the second quarter of the nineteenth century, the Maronites were already closely allied with the French and had emerged as the most important local sectarian power. This brought them into conflict not only with Ottoman authorities, but more immediately with the Druze and other groups in the area. In 1843, following a series of uprisings in Lebanon, the Ottomans responded to European pressure to create separate sectarian governates for the various groups they ruled. Following a series of massacres between the Druze, supported by the British, and the Maronites, supported by the French, the Ottomans created in 1864 yet another governmental apparatus, one which is essentially the basis for the present state of Lebanon. This was the Governate of Lebanon, based on a system of sectarian representation, with an appointed Maronite governor. The political and economic dominance of the Maronite community institutionalized at that time continued until tested and broken in the 1975 civil war in Lebanon.[11]

The Republic of Lebanon was proclaimed in 1926 when the Constitution was promulgated, but the real reins of power remained with the French. In 1943 the various Lebanese political leaders and factions closed ranks and demanded independence. When the French ousted the president and his prime minister, the Lebanese formed a resistance

[11] Albert Hourani, *Syria and Lebanon; A Political Essay* (London: Oxford University Press, 1946). See also Albert Hourani, "Lebanon: The Development of a Political Society," in *Politics in Lebanon*, ed. Leonard Binder (New York: John Wiley, 1966).

government and proclaimed a National Pact, which regulated relations between the different communities, of which the Maronite was then the largest. Complete independence was not achieved until the end of 1946, when the last French soldier sailed away and Lebanon became a member of the Arab League and the United Nations.

Modern Lebanon was thus founded on the assumption that the different confessional groups would be united in a single society as corporate units holding equal rights and status in the public domain. Thus citizenship in Lebanon came to necessitate identification with a religious group, the seven main ones being: Maronites, Sunni Moslems, Shi'a Moslems (the largest single sect), Greek Orthodox, Greek Catholic, Druze, and Armenian Orthodox (Gregorian). Other lesser groupings included Jacobite Christians, Syrian Catholics, Armenian Catholics, Jews, and Protestants—all this in a population of little more than 3 million.

The different confessional groups in Lebanon shared a common language, Arabic, and a recognition of their interdependence. No community could dominate the other without endangering the very existence of the state, which was maintained in a "precarious balance," or perhaps more aptly, in the words of the London Times, "in a balance of fear." Once hailed as the model for a Middle Eastern pluralistic society, Lebanon served a key function in the area. In the past, it was a refuge for persecuted religious minorities, and in modern times it became a haven for Middle Eastern political refugees of all persuasions.

And then in the summer of 1975, the "precarious balance" came to an abrupt end. The delicate political fabric of Lebanon began to unravel and civil war broke out, overtly pitching Christian against Moslem in a war whose aftermath is still to come. Of course, it is difficult and foolhardy to analyze a situation in the making, and even at the time of this writing, the strife in Lebanon is far from over. However, a few clarifying remarks are necessary in order to correct the overly simplistic interpretation of the "Christian-Moslem" conflict in Lebanon as being no more than an inevitable ethnic confrontation. As with most ethnic confrontations, the one in Lebanon is symptomatic of profound socioeconomic dislocations, here greatly exacerbated by outside forces.

It has been noted quite accurately that the catalyst for the civil war was the presence in Lebanon of a large Palestinian population, both as refugees in camps that stretched from the heart of Beirut to the southern borders and as an armed military group that constituted a state within a state. The presence of the refugees put a serious strain on Lebanon's economy and the well-armed Palestinian militia complicated the political scene both internally and at the international level. Their presence in and use of south Lebanon as a base of operations prompted regular and severe retaliatory strikes by Israeli forces. The effect of these strikes reverberated throughout the Lebanese society and further amplified the strains that existed among the different groups.[12]

The presence of the Palestinians clearly precipitated the crisis, but there are other underlying factors that also contributed. One factor that affected the subsequent conflict has to do with the way that sectarian and ethnic groups are distributed throughout the country. Each of the seventeen officially recognized sectarian or ethnic groups in Lebanon has a relatively clear regional base from which its leading families contend for power on the national scene.[13] A second underlying cause of the ongoing conflict has to do with differential access to the resources available for each group. The prosperity the country experienced in the 1940s and 1950s as a result of its role as a leading trade and banking center was unequally shared among the different groups; some, like the Shi'a, were substantially excluded from benefiting from the economic growth that favored most of the Christian communities engaged in trade.[14]

Prior to World War II, it seems that economic disparities among the various sectarian communities were relatively muted and were channeled along well-established and fairly acceptable lines. By the mid-1970s, however, the agricultural majority of the population was receiving only 15 percent of the Gross National Product, while the 14 percent of the population engaged in commerce and related activities received 46 percent of the GNP. The Maronites benefited disproportionately from this development. Also, although no census has been conducted since 1932, the belief was widespread that the Maronites no longer constituted the largest sect in the country, and that their political power, while reflecting wealth, exceeded their numbers, giving another cause for resentment.

In earlier years sectarian identification held little implication for economic class differentiation; by 1975 such was no longer the case. In February 1975 there was a major strike by Shi'a fishermen working for Maronite boatowners; these violent protests were quickly joined by Palestinians and others.[15] With their numbers considerably enhanced by

[12] Laurel Mailloux, "Peasants and Social Protest: 1975–1976 Lebanese Civil War," in *Muslim-Christian Conflicts: Economic, Political and Social Origins*, eds. Suad Joseph and Barbara Pillsbury (Boulder, Col.: Westview Press, 1978), pp. 99–128.

[13] For a discussion of the Lebanese pattern of "big man" leadership, see Michael Hudson, *The Precarious Republic: Political Modernization in Lebanon*, (New York: Random House, 1968); see also Samir Khalaf, "Primordial Ties and Politics in Lebanon," *Middle Eastern Studies*, 4, no. 3 (1968), pp. 243–69.

[14] Suad Joseph, "Muslim-Christian Conflict in Lebanon: A Perspective on the Evolution of Sectarianism," in *Muslim-Christian Conflicts*, pp. 70–78.

[15] Mailloux, "Peasants and Social Protest," p. 115.

the Palestinian refugees, the various Moslem groupings demanded a larger share of the economy and a reorganization of the political system, with increased secularization of the state and mass elections. The dominant Christians, fearful of losing their advantage and angered by the alien presence of the revolutionary Palestinians—Christian or Moslem—refused a drastic overhaul of the political system and armed confrontation ensued. But it was quickly apparent that Lebanon was to become the arena for the power plays of several outside interests: Syrian, Israeli, and others. The instability generated by outside intervention and acts of intercommunal violence quickly underminded the already weak central state institutions and initiated a period of civil war.

The alignments to which continuing warfare has given birth are bewildering. Even close observers disagree as to the underlying principles that govern the short-term alliances among the many contending parties: Shi'a ally with Palestinians; right-wing Phalangist Christians are (or were) supported by the Syrian Army; Israel aids certain Christian armed militia in the south. Many who rally supporters to their cause in terms of religious or political beliefs appear to have little more than banditry in mind. It is a conflict whose destruction has clearly gone beyond the self-interest of any class or sect. If any interest is served in all this, it is certainly not that of the Lebanese.

THE ARMENIANS

Another Christian community in Lebanon is that of the 180,000 Armenians who live mainly in the Burj Hammoud area of Beirut. While the majority belong to the Armenian Orthodox Church, small groups are Roman Catholic and Protestant. The Armenians were granted Lebanese citizenship in 1939 under the French mandatory regime. The Armenian population of Lebanon and the approximately 70,000 Armenian citizens of Turkey, resident for the most part in Istanbul, are part of a larger diaspora which includes approximately 4 million in the U.S.S.R., 1 million in the U.S. and Canada, and around 400,000 in France.

The Armenian community was formerly the most important and influential *millet* of the Ottoman Empire, and through its long history provided counselors to the Ottoman court and contributed actively to the commercial and intellectual life of the Empire. The total Armenian population in Anatolia at the turn of the century has been estimated at about 1,800,000, with the majority living in villages in the central and eastern regions. In the steppes and mountains of Anatolia, Armenian villages were intermingled with those of Kurdish, Turkmen, and other

Moslem and Christian groups. The *millet* itself was traditionally headed by a Patriarch, confirmed by the Ottoman Sultan, and generally selected by influential clerics and a number of wealthy urban families.

In the nineteenth century, the Armenian *millet* was subject to the same economic and political changes that buffeted the rest of Ottoman society, including the rise of nationalism which had sparked the separatist movements of the peninsular Greeks and Balkan Christians. Caught between the British-Russian rivalry and the desire of many Armenians for internal reform, the *millet* became faction ridden and highly politicized. The traditional leadership was opposed by those who, calling upon the support of one or another European power, attempted to secure further political advantage or even establish an independent Armenia. Influenced by the French Revolution and especially by the 1830 and 1848 revolutionary movements in Europe, young Armenian activists founded two socialist revolutionary parties, the Hunchakian and the Dashnak. The first, founded in 1887 in Geneva, called for complete political independence; the second, founded in Tiflis in 1890, advocated reforms within the framework of the Ottoman Empire.

Reacting violently to the threat of the Armenian revolutionary movements, a number of Armenian communities were attacked. The period 1894 through 1916 was marked by much intersectarian strife, repeated attacks on Armenian villages, and anti-Armenian pogroms. European powers consistently intervened in Ottoman domestic affairs and this, by all accounts, contributed further to the widening gap that separated Moslem and Christian communities. French, British, and Russian consulates regularly issued passports to Turkish Christian subjects, supplied money to dissidents, and encouraged the nationalist aspirations of the Greeks and the Armenians.

During World War I and immediately after the breakup of the Ottoman Empire, massive deportations, forced movements of populations, and even the starvation of substantial numbers ultimately resulted in the removal of nearly all Armenians from rural Anatolia. The ill-considered attempt to establish an Armenian Republic in southeastern Turkey following World War I resulted in the removal of the last remaining rural concentration of Armenians from Turkey. Thousands fled to Lebanon and Syria. Following World War II, about 40,000 Armenians left Lebanon and Syria and emigrated to Soviet Armenia, but the rest remained in Lebanon, where today they form a strong, internally cohesive ethnic community under the leadership of the Dashnak party.[16]

[16] For a succinct discussion of the transformation of the Armenian community in Beirut, see Suad Joseph, "Transformation of the Armenian Movement in Lebanon," unpublished manuscript, n.d.

MOSLEM-DERIVED SECTARIAN COMMUNITIES

Although more could be said about the nature of non-Moslem communities and their constituent social groupings, in almost every country of the Middle East we have seen something of a shared pattern. This is one of encapsulation rather than assimilation of communities whose outer limits are set by religion. As described in Chapters 2 and 3, there are numerous Moslem-derived communities as well. In a structural sense, they are very similar to the non-Moslem ones, forming as they do inward-looking confessional minorities. Two of the latter mentioned earlier are the Druze of Syria, Lebanon, and Israel, and the Alawis (Alevis) of Syria and Turkey.

Groups like the Druze and the Alawis exemplify the use of distinctive Islamic ideology and practice to announce a separate identity and to maintain what amounts to closed communities within the larger society.[17] What these and others like the 'Ibadis of Oman and the Metwali Shi'a communities of Lebanon and Syria share is a history of political dissidence and persecution. Followers of these movements survived as weak minorities at the periphery of Islamic society; all sought refuge in rural, economically marginal, hard-to-administer areas, or "refuge zones." As a consequence, the political and social life of these communities tends to be highly involuted, inaccessible to outsiders, and hedged with secrecy. Leadership is usually in the hands of an oligarchy of religious leaders or elders.

Only recently has the spread of nationalism and the intrusion of state-wide institutions, particularly public education, begun to erode some of the cultural boundaries that separate these sectarian groupings. In Turkey, for example, prior to the 1980 junta, some of the Alawis participated in leftist party politics as they sought a share of political power and economic gain. In Syria the once poor and isolated Alawis or Nusairis, through their disproportionate representation in the army, managed to achieve a near monopoly in the leadership of the country.

One case of religious dissidence whose outcome is virtually unique within recent Islamic history is the Baha'i movement. This movement originated in Shiraz, Iran, in 1844, when a young man proclaimed himself the *Bab*, or "Gateway to Heaven," and the new manifestation of the Prophet Muhammad. He rapidly gathered a following as he preached against the corruption of the clerical and governmental establishment of the Persia of his day. In this, the movement he founded followed a familiar pattern of expressing political and social protest in the idiom of religious reform. However, once Babism, as it was called, was put down

[17] For a description of the Alevis in eastern Turkey, see Nur Yalman, "Islamic Reform and the Mystic Tradition in Eastern Turkey," *European Journal of Sociology*, 10 (1969), 41–69.

by the execution of its leader in 1850 and the brutal persecution of his followers, the movement itself was radically transformed. Baha'allah, half-brother of the founder, began to interpret Babism as a universalistic faith, trying to reconcile what he perceived to be the best of Judaism and Christianity with Islam. Ultimately the faith he founded, Baha'ism, broke with Islam. Today its followers are found throughout the world. One of their important temples, for example, is in Evanston, Illinois, while the spiritual leader, a descendant of Baha'allah, resided in Haifa, Israel.

Today most of the small Baha'i community of the Middle East is found In Iran, where, until recently, they were relatively well-to-do, were involved in business and commerce, and reputedly had close connections to the Shah's family. Now, following the Revolution, they are again suffering persecution and discrimination. Many have recently been executed. What distinguishes the Baha'i movement is that it resulted in what amounts to a new universalistic religion, one that has neither clergy nor elaborate ceremonies and that emphasizes the principles of humanitarianism and pacifism.

REGIONAL ETHNIC GROUPINGS

We have stressed dissidence within Islam and non-Moslem sectarianism as important sources of ethnic and cultural diversity. In addition to sectarian-defined groupings, there are important Moslem populations

Turkmen farmers going to market in northeastern Iran.

that are distinguished by several overlapping claims to shared identity. Foremost among these are the shared sense of history, language, and cultural heritage associated with a region or place of origin. Such groups may range from large populations with strong territorial bases, like the Baluch, the Kurds, and the Palestinians, to smaller dispersed ones, like the Circassians. The Circassians, Moslems who fled the Caucasus in the nineteenth century, are found today on the Golan Heights, in Jordan, Syria, Iraq, and Turkey. As a small and widely dispersed people, they appear to aspire to no more of a political future than to participate in the national order of the countries in which they are found. Today, the Kurds, Baluch, Palestinians, and the Azeris are divided among a number of different nation-states. Given their size and their long-established historical presence in their homeland, these people can be thought of as incipient nations without states. This is also perhaps the case for the Turkmen of north-central Iran, Afghanistan, and the Soviet Union. Thus it is not surprising that today all of the just-mentioned groups have more or less active nationalistic movements seeking political expression either in independent nation-states or as recognized entities in confederate states.

CHAPTER FIVE PASTORALISM AND NOMADIC SOCIETY

Nomadic pastoralists are people pursuing a way of life which, perhaps more than anything else, seems to typify the Middle East in the eyes of Westerners. The image of the mounted tribesman of the desert is all too often taken as a metaphor for all Middle Eastern peoples, perhaps as the cowboy is sometimes said to embody social values and an ethos particularly American. Reality is understandably more complex. Although the Middle East is one of the few areas of the world where substantial numbers of people have lived by specializing in animal production, agriculture is much more important by far. Pastoralists account for a small percentage of all food produced, and in most regions they do not even supply most of the meat and dairy products consumed.

Still, nomadic pastoralism is supportive of a distinctive cultural tradition and way of life for many. More important, tribally organized nomadic pastoral peoples have been politically and historically significant beyond their numbers. We already saw some of the reasons for this when we sketched the events surrounding the rise and spread of early Islam. Although by no means can we speak of the early Arab Empires and city-states as nomadic—as unfortunately some are all too prone to do—we do know that the mobility and military prowess of Arabian tribes played a significant role in early conquests of Byzantine and Persian provinces. Further, we know that the importance of Mecca as a mercantile center rested in large part on its ability to control and utilize nomadic tribes for the operation and protection of long-distance trade routes.

Who, then, are the nomadic pastoralists of the Middle East? Why do they occupy such a distinctive place in the fabric of Middle Eastern society? What problems do such peoples face in the latter half of the twentieth century? And what are their prospects? To answer these and similar questions we will look at a number of populations including the Yörük of Turkey, the Basseri of Iran, and the Murra of Saudi Arabia. None of these are "typical" in the sense that they represent fixed patterns or can be taken as stereotypes. Rather, each group is a contemporary population, facing a range of unique circumstances in attempting to solve problems common to all peoples.

In some respects, the nomads of the Middle East appear to represent the survival into the industrial age of an ancient and unchanging way of life. On the surface this is true enough, because some of the people we discuss continue to live in woven goat-hair tents little changed for centuries. Their annual routine of seasonal movement in search of grazing for their animals is as old a phenomenon as agriculture itself. The tent groups and the tribal idiom of social organization follow principles that are described by the early chronicles of Islamic history. But as we shall see, visitors today fortunate enough to be taken into the homes of contemporary pastoralists do not find relics of the past, but families who are very much at home in today's world, coping with many of the problems of contemporary society while attempting to enjoy its obvious material advantages. The Toyota truck parked outside the family tent is far more common a sight today than is the mounted herdsman. Before turning to specific people and places, let us review some of the more general observations that can be made about pastoralism in a region as complex and varied as the Middle East.[1]

THE ECOLOGICAL BASIS FOR PASTORALISM

So far we have spoken of nomadic pastoralism as if it represented one uniform cultural tradition or way of life. In fact, although they share certain characteristics, nomadic pastoral societies are no more like one another than are communities engaged in agriculture. Every population

[1] Three important collections of both research-oriented and theoretical articles on pastoralism are *Comparative Studies of Nomadism and Pastoralism* in a special issue of *Anthropological Quarterly*, 44, no. 3 (July 1971); William Irons and Neville Dyson-Hudson, eds., *Perspectives on Nomadism* (Leiden, Brill, 1972); and *Pastoral Production and Society*,

has a unique history and cultural heritage including language, sectarian affiliation, and customary ways of behaving. In southeastern Anatolia, Turkish-, Arabic-, and Kurdish-speaking nomadic and sedentary groups share a common area. Likewise, in southwestern Iran a large multiethnic tribal confederation, the Khamseh, or "Five," encompass Persian-, Turkish-, and Arabic-speaking farmers and nomadic herders. Even among groups who share a language and a common general designation, there may well be important sectarian or religious divisions. The Arabic-speaking tribes of Khuzistan divide along lines of Sunni-Shi'a affiliations, as do some of the Kurdish groupings. Even with the major Turkmen tribes of north-central Iran, there is a fundamental distinction between those sections who claim sacred descent and those who do not.

The great cultural diversity among nomadic groups makes it prudent to beware of easy generalizations about nomads and their life. There are, however, a number of variables relating to pastoralism as a means of production that select for certain regularities in social and economic organization. Important variables include the degree and types of movement which are incorporated into the system of livestock management. Other variables are related to the nature of the animals kept, their food and water requirements, and the labor it takes to herd them. Also important is how the animals are converted into the foods that the family eats and the items that they purchase—in other words, how the people make use of their animals within a larger system of exchange and trade.

The Middle Eastern landscape can be divided into three major zones of land-use potential. The first and the most restricted in distribution is land suited to village agriculture and urban settlement. Here, as we have already noted, water is the critical variable. The second zone comprises lands almost totally unfit for human habitation—for example, the Dasht-i Kavir and the Lut deserts of Iran, and the Rub' al-Khali of Arabia. The third and largest zone comprises land marginal to agriculture by virtue of aridity or altitude. This zone, however, can be exploited with great investment of labor or on a limited seasonal basis. The desert areas support natural grasses and plants on a seasonal basis, as do the mountain slopes and high valleys and the extensive rugged steppes of Iran, Afghanistan, Iraq, and Turkey. It is in such marginal areas that pastoralism is a practical alternative or supplement to farming. Although most farmers in both the well-watered and marginal areas keep livestock for domestic use or even for sale, very little specialized livestock raising is engaged in by sedentary peoples. This is, of course, an important contrast between Europe and the Middle East. In Europe,

---

Proceedings of the International Meeting on Nomadic Pastoralism, Paris, December 1–3, 1976 (Cambridge, Eng.: Cambridge University Press, 1979).

as in the United States, specialized dairy farming and sedentary stock raising are important forms of land use.

The reasons why dairy farming and ranching have not evolved in the Middle East are largely ecological and economic. In an arid area supporting a relatively high density of population and large concentrations of people in urban centers, it is a costly luxury to use arable land and grasses primarily to raise animals. It is far more efficient to raise cereal crops for human consumption than to attempt to convert grains and grasses into animal products. The diet in the Middle East reflects this. There is a strong emphasis on bread, together with dishes using vegetables and grains. Meat is used sparingly, usually for broth and stews.

Animal husbandry as a major or specialized strategy pays off economically only when it increases net food production. This is usually in areas where animals can be grazed on land too dry, too uneven, or too high in altitude for regular cultivation. As a consequence, grazing areas are almost invariably tracts of wild growing plants where the available forage is limited. The animals must be moved regularly from pasture to pasture, often migrating over a grazing range of several hundred kilometers annually. Sometimes farmers and nomads effect a trade-off that benefits both.

In areas where farming is possible but long fallow periods are necessary to maintain soil fertility, as was traditionally the case along the Mediterranean coast and on the Syrian steppes, nomadic pastoralists may adjust their migratory schedules in order to graze their flocks on fallow fields. The advantage to their field owners is that the animals supply valuable fertilizer. This arrangement, though potentially mutual in its benefits, is by no means without serious conflict of interest. Animals often damage crops and may encroach on land under cultivation. Also villagers themselves may raise sufficient animals to fully exploit available local grazing, and nomadic pastoralists may represent a direct economic threat. Today the governments of the Middle East carefully regulate such peasant-nomad interaction as exists. In Turkey, for example, nomads very often may pay rental fees in order to use village lands. In Syria the government closely controls nomadic grazing and migratory routes, as did the Iranians with regard to most of the larger nomadic tribes.

ANIMALS AND NOMADISM

One convenient way to summarize the diversity in nomadic movement and land use is in terms of two contrasting patterns: transhumance, or vertical migration, and plains, or horizontal migration. *Transhumance* is a pattern of animal management found in mountain areas throughout the world. It involves seasonal movement of herds between different climatic zones. In spring the road and trails of the Zagros, Taurus, and Elburz ranges become crowded with flocks, as the nomads and villagers move their herds from the lowland winter pastures up to the mountains. In the fall the pattern of migration is reversed, as the herds are led down the mountain ranges to spend the winter months in a warmer zone. Thus, transhumant nomads usually have two major grazing areas, and their tents may remain in one place for the duration of the long grazing seasons. As is common in Kurdistan, some tribes may have permanent homes in the villages they occupy during the winter season. Although cattle are herded in some parts of Taurus and Zagros, the majority of Middle Eastern transhumants specialize in sheep and goats. Today the basic pattern of movement persists, but most animals are transported by truck between the pastures.

Horizontal nomadism usually involves a greater movement of both herds and households. This is because different pasture zones are apt to be further apart than they would be in a mountainous area. Water, and not simply the availability of pasture, is the determinant of how many animals can be managed and where people camp. The greater distances traversed in this type of nomadism foster a greater reliance on large transport animals like camels. Nomads utilizing camels and horses have historically played an important military role in the areas where they are found. Well mounted and knowing their terrain, they could frequently maintain a military advantage in conflict even with government troops. Further, their tribal organization facilitated territorial defense and allowed them to maintain partial, if not complete, autonomy. Even today, the Turkish-speaking Qashqa'i confederation of southwestern Iran, politically suppressed by the Shah, are again reasserting their local claims to power. Many sedentarized nomads are back on their migration and as tribal groups are reclaiming their lands.[2]

The type of animal or mix of species herded by pastoralists varies with the terrain and with market demands. Herded livestock includes camels, cattle, sheep, and goats. Although we often think of the camel when we speak of the Arabian Peninsula, in fact even there, with the exception of the particularly arid wastes, the most commonly herded animals are sheep and goats. Cattle are fairly common in the extreme southern reaches of the Peninsula, as well as throughout the mountainous areas of the Middle East. In Iran a few groups, such as the Baluch, raise large numbers of camels, but for the most part sheep and goats are the mainstays of nomadic pastoral life. In highland eastern

[2] Lois Beck, "Herd Owners and Herd Shepherds: The Qashqa'i of Iran." *Ethnology*, 19, no. 3 (1980), 345–52.

Syrian Bedouins with their herds. (Photo courtesy of Ulku Bates)

Turkey and Iran, some Kurdish tribes have specialized in cattle, but again the preponderance of pastoral production involves small animals who are well suited to browsing in rough, broken country.

In the arid reaches of Arabia on the margins of Rub' al-Khali, in the Gulf states, and along the southern rim of the Syrian steppe camels were long the key to human adaptation. There the extremely rough forage, great temperature variability, and long droughts, together with the need to cover long distances between water holes, made this slow-maturing animal supremely suited to the desert habitat of the Bedouin. The term *Bedouin,* or more correctly *Bedu,* is the Arabic word applied to the nomadic tribes of the desert. It sometimes takes on a more restricted connotation as herders of camels.

The Bedouins traditionally raise camels to the near exclusion of all other animals, except for their prized racing and riding horses. Sheep herders, commonly called *shawiyya,* generally keep few camels. Both groups reside year round in black tents woven from goat hair; they follow different patterns of migration and are separated by a considerable social gulf. Camel herders consider themselves higher in rank than sheep herders, and, among themselves, some lineages are considered to be more "pure" and noble than the rest. These core lineages have traditionally provided the leaders or paramount sheikhs of the different tribes and confederations.

During the hot summer months of June through September, when the camels must be watered nearly daily, camel herders by necessity cluster around wells and other permanent sources of water. In the other months, when moisture is available from the annual plants, the camels can go long periods without being watered; in winter camels are not watered at all, and the nomads range far into the arid zone.

Those sheep-herding *shawiyya* who are still nomadic camp on the edge of the desert and group around wells in the summer; in winter they take their flocks through village lands, using fallow fields made available to them after the fall harvest. Traditional dependence on sheep and goats used to limit their mobility, for these animals must be watered regularly in all seasons and are incapable of moving long distances at any appreciable speed. Today this has changed dramatically, as most use trucks to reach even distant grazing areas far outside their customary ranges. Changed also is the old pattern whereby the Bedu of the interior extracted tribute or "taxes" from their less-powerful neighbors.

In the past, camel-breeding Bedouin tribes of the Arabian Peninsula conformed more to our image of the "desert warrior" in that martial arts were highly regarded. The celebrated raids (*ghazzu*) carried out against the camps of other nomads or against villagers earned them a formidable reputation. The fact is that for the larger tribes raiding and warfare constituted an important part of their adaptation.[3] Many tribal sections controlled tributary settled populations, exacted protection money (*khuwwa*) from border villages, and received regular payments for the supervision and protection of trade routes. This latter often meant simply supplying guides and agreeing not to attack passing caravans. Some nomadic tribes of the desert kept slaves who were attached to the households of the wealthy.

PASTORALISM AND MARKET RELATIONS

The camel herders, however much they may prize their independence, are still closely linked by numerous economic and social ties to the larger society. This is true for all pastoralists, whether they live and raise their animals in close proximity to villages and towns, or whether they range far into regions of no permanent settlement. Pastoralists in the Middle East raise animals for sale in markets, where they acquire most of the food they consume, not to mention clothing, cooking utensils, weapons, jewelry, and the many items of household and personal use. The integration of the nomads into the larger market economy is a point

[3] See William Irons, "Livestock Raiding among Pastoralists: An Adaptive Interpretation," *Papers of the Michigan Academy of Science, Arts and Letters*, vol. 50 (1965); and Louise Sweet, "Camel Raiding of North Arabian Bedouin: A Mechanism of Ecological Adaptation," *American Anthropologist*, 67, no. 4, (1965), 1152–50.

that cannot be emphasized too strongly. In fact, the existence of specialized pastoralism and the number of people who can be supported in this endeavor at any given time are directly related to the ability of pastoralists to exchange for or acquire the products of sedentary communities.[4] Moreover these exchanges are effected, for the most part, through cash transactions.

The diet and general consumption patterns of the nomads are not substantially different from those of village dwellers. Their diet is basically the same as that of any other rural family, with perhaps a higher consumption of meat and other animal products. Bread is the staple almost everywhere, and to acquire wheat or barley flour a family must regularly sell its animals or herd products. No household in the Middle East is truly self-sufficient, and few are more specialized and hence dependent than are nomadic pastoralists.

Even before the advent of trucks, Bedouin camps in Arabia were regularly visited by two types of merchants. One class of shopkeepers, called the *Kubaisat* after a town in south Iraq, visited winter encampments in the desert with camels laden with such goods as cloth, sugar, rice, and small household objects. Their shop was set up in a white tent—to distinguish it from the black tents of the nomads—and to ensure that the merchant would not be mistakenly attacked in a sudden raid. It was considered a serious breach of the codes regulating desert warfare to assault or even rob such nontribal traders, so important was their function to the survival of the nomad pastoralists. The *Kubaisats* paid a set fee to the sheikh or leader of the tent groups for the privilege of joining the camp and being under the leader's protection.[5]

The second type of traveling merchant in the Arabian desert were the members of the 'Aqail tribe, who acted as agents for mercantile houses in Cairo, Damascus, Basra, and elsewhere. These agents bought camels, branded them, and took them to urban markets. Again, these merchants were protected by the established codes of war, and in the event their camels were taken in a raid, they would be returned.

Today, of course, this situation is very different, as Donald Cole, Emanuel Marx, Dawn Chatty, and other students of Bedouin life note.[6] Some households within each camp have jeeps or trucks with which

[4]Daniel Bates and Susan Lees, "The Role of Exchange in Productive Specialization," *American Anthropologist*, 79 (1977), 824–41.

[5]Carleton Coon, *Caravan: The Story of the Middle East* (New York: Holt, Rinehart Winston, 1958), p. 191.

[6]Donald Cole, *Nomads of the Nomads: The Al Murrah Bedouin of the Empty Quarter* (Chicago: Aldine, 1975); Emanuel Marx, *Bedouin of the Negev* (Manchester: Manchester University Press, 1967); and Dawn Chatty "Changing Sex Roles in Bedouin Society in Syria and Lebanon," in *Women in the Muslim World*, eds. Lois Beck and Nikki Keddi (Cambridge, Mass.: Harvard University Press, 1978), pp. 399–415.

they commute regularly to market towns. Among the Bedouins mechanized transport has almost totally replaced reliance on the camel, and markets are readily accessible on a daily basis. The situation is very similar among the other pastoralists of the region. In many countries, pastoralists, like ranchers in the United States, follow livestock prices closely and go to town to consult with livestock brokers and to learn the latest prices for animals. Rapid road transport gives pastoralists a choice of markets and ready access to the consumer goods of the city.

A frequent scene in any market town is that of the herdsman transacting business with merchants and artisans. Because each household consumes several hundred kilos of grain a year—not to mention tea, tobacco, sugar, and clothing—these transactions may involve substantial sums of money. Business is generally conducted with elaborate courtesy and with mutual signs of respect. Merchant-customer relations often include the extension of credit from one season to the next, and such relationships of mutual trust may be passed on from one generation to another.

Pastoral households, perhaps more so than their peasant counterparts, are directly dependent on others for the food they eat. This renders them particularly vulnerable to changing market prices of animal products and forces many of them to seek alternative sources of food or income. They thus often engage in trade, part-time farming, and other pursuits such as seasonal labor in cities or on farms. All in all, the vagaries of the marketplace are as critical as are those of the weather.

Philip Salzman, who has worked extensively with the Yarahmozadi Baluch of Iran, says that "multiple-resourcefulness" would be a more appropriate characterization of the Baluch adaptation—indeed, if not of most pastoralists. The Baluch migratory cycle is determined only in part by the needs of their animals. They also move on schedule as they hire themselves to help in harvesting dates or tilling fields, and even to engage in smuggling.[7]

POLITICS AND NOMADIC PASTORALISM

Although so far we have stressed the economic aspects of pastoral adaptation, there is more involved. No way of life is the outcome of strictly economic pressures. Every community lives in close proximity with other groups, sometimes with those on whom it directly depends for a critical resource or item of technology. Just as often, the

[7] Philip Salzman, "Multi-Resource Nomadism in Iranian Baluchistan," in *Perspectives on Nomadism*, eds. William Irons and Neville Dyson-Hudson (Leiden: Brill, 1972), pp. 60–68.

sometimes A
neighboring group is a real or potential competitor or even an enemy. The organization of any nomadic society and even its pattern of movement are also responses to its political environment—namely, to adjacent populations and to the power of the encompassing state.

States or central governments in the Middle East are often critical in determining local patterns of land use and how local communities interact, even in areas remote from centers of power. A government may not be able to effectively administer certain territories within its national frontiers on a day-to-day basis, but nevertheless it can usually intervene in local disputes in such a way as to determine their outcome. Because nomads are most commonly found in regions of low population density, remote from urban centers and in terrain difficult to control militarily, it is sometimes assumed that national politics are irrelevant. This is not the case either historically or today.

Modern technology has, of course, greatly reduced the military advantage of mobility that nomads have traditionally enjoyed; this is the case even in Iran, where the central government is now relatively weak. Modern governments take care to maintain their monopoly over military armament, and even where tribal groups may be well armed, they are usually vulnerable to air attack and to the ability of the government to bring in large numbers of troops by truck. Such large tribal groups as the Basseri, the Turkmen, and the Qashqa'i are far from being autonomous political entities. However, even when such nomadic groups are able to keep the bureaucracy at arm's length, they still have to contend with a state that can play off one leader against another, interfere in factional disputes, and, on occasion, launch large-scale punitive expeditions.

William Irons, who has worked extensively among the Yomut Turkmen of northeastern Iran, goes so far as to say that their nomadism was largely a political response to the Iranian government's attempt to establish control over their territories. The Yomut moved their *obas,* or camps, not simply to maintain their flocks, but to enhance their military capabilities vis-a-vis the troops of the Shah.[8]

The interaction of people and groups as they struggle for scarce resources, for power, or even for such elusive social rewards as prestige can be as important as the dictates of a mode of production in shaping how people will use the lands they inhabit. One case study, the Yörük pastoralists of southeastern Turkey, offers a good illustration of how

[8] William Irons, *The Yomut Turkmen: A Study of Social Organization among a Central Asian Turkic Speaking Population,* Museum of Anthropology, University of Michigan Anthropological Papers, no. 58 (Ann Arbor: University of Michigan, 1975); also William Irons, "Nomadism as a Political Adaptation: The Case of the Yomut Turkmen," *American Enthnologist, 1,* no. 4 (1974), 635–58. For a recent book-length account of politics and land use among Iranian pastorals, see Richard Tapper, *Pasture and Politics,* (New York: Academic Press, 1979).

changes in intergroup relations affect internal organization as well as regional patterns of land use. After a brief sketch of the modern history of the Yörük, we can begin to draw a more detailed portrait of them as people.[9]

THE YÖRÜK OF SOUTHEAST TURKEY

In the Middle East, population movements have been such that we cannot take for granted the local antiquity of any particular group. The Yörük are relative newcomers to the Gaziantep area near the Syrian border, where they are presently the only people who specialize in animal husbandry. Although not all tribal sections arrived in the region at the same time, it is clear that none were established there before 1900. Furthermore, even though groups came to the southeast from various parts of Anatolia, all originated on the west coast, from where they dispersed in the late nineteenth and early twentieth centuries. Thus, to consider why the Yörük are where they are today may be a useful way to see some of the problems they, and many other groups as well, faced and how they coped with them.

The Yörük, a Turkish-speaking people, were traditionally nomadic goat herders along the Aegean and western Mediterranean coasts of Turkey. Although it is not clear when they emerged as a distinct ethnic or tribal group, they are mentioned by name in twelfth-century texts The Yörük themselves trace their origins to northern Iran and regard themselves as descended from the Turkmen. As such, they view themselves as among the earliest Turkish settlers in Anatolia.

The Yörük moved seasonally to exploit zonal and seasonal variation in wild grasses in the mountain range of western Anatolia, and thereby maintained herds in numbers beyond that which could be supported in any one place throughout the year. Moving in small camp groups, they would winter in the coastal plains, grazing their flocks on fallow or nonarable land. In the spring, in the classic transhumant pattern, the herds would be taken up to successively higher grazing areas. By early summer they reached the upper pastures, called *yayla*, of the Taurus range, where they would camp until fall. At the onset of cold weather they would retrace their route downward. By then villages on the middle and lower slopes of the seaward side of the Taurus would be finished with the grain harvest, and Yörük flocks could pass without risk

[9]Daniel Bates, *Nomads and Farmers: A Study of the Yörük of Southeastern Turkey*, Museum of Anthropology, University of Michigan Anthropological Papers No. 52 (Ann Arbor: University of Michigan, 1973).

Yörük woman and child in a migration in 1969. Today most households move by truck.

of crop damage, grazing on the stubble while they enriched the fields with their droppings.[10] The Yörük occupied a specialized niche and constituted one group among several populations engaged in different forms of land use.

The migratory schedule connected grazing areas that were seasonally available without the Yörük's being subject to much direct competition with other groups. Thus the place of the Yörük in the regional economy was a product of different populations interacting with each other. In fact, as John Kolars, a geographer, points out, this pattern of interaction was beneficial to all concerned. The Yörük used land that villagers could not exploit, and the nomads' flocks supplied peasants with pastoral products.

[10]Xavier de Planhol, "De la plaine pamphylienne aux lacs pisidien: Nomadisme et vie paysanne," Bibliothèque Archéologique et Historique de l'Institut Français d'Archéologie d'Istanbul, tome 111 (Paris: Librarie Adrien-Maisonneuve, 1958).

In the late nineteenth and early twentieth centuries, the western coast of Turkey became the focus for intensive agricultural development spurred by European and domestic investment and the opening of a new rail link. As the marshes were drained, areas that were formerly winter pastures for the Yörük were settled by new villages, often set up or encouraged by the government. The strategies available to the Yörük for managing their herds became more restricted. Not only was there direct competition for scarce resources, but also for the balance of power among groups, as some were now supported by the government.

As a result of these pressures, particularly because of diminished pasture, Yörük families were obliged to alter their previous patterns of herding and migration and even how they made a living. A substantial number of pastoral families chose to take up farming and settle in villages. Others shifted their migratory routes. Some moved to less developed but poorer areas of grazing on the coast, some to central Anatolia, and still others to southeastern Turkey. The latter constitutes the only large concentration of Yörük households that are still nomadic, as virtually all the remaining Yörük populations today are settled. Like many ethnic or tribal groupings in the Middle East, the Yörük are widely dispersed.

Those Yörük who migrated to the southeast were able to continue as nomadic pastoralists only because other nomadic populations had been forced to abandon pastures in the lowlands and high mountain valleys. By then the government had forcibly settled local nomadic Kurdish, Arab, and Turkmen tribes. The Yörük were permitted to enter the region because they were not considered a political threat, as were the larger and more powerful Kurdish and Turkmen groups.

The Yörük families who continued to be pastoralists soon shifted from goat to sheep production and established an annual routine of migration that persists today. The change in animals herded was due to governmental control of routes to mountain pastures. As these passed through forested areas, access to them was prohibited to goats in an effort to slow the rate of deforestation which threatened still magnificent stands of pine and hardwoods.

The shift in emphasis from goat herding to sheep herding was not the only change in Yörük life. In 1949 and 1950 the Yörük, like other rural populations in Turkey, were greatly affected by specific national economic and political developments. A new government, concerned with rural investment and agricultural development, greatly increased the availability of agricultural credit. All-weather roads were extended into the areas of winter pasture, and the resultant expansion of agriculture caused a decline in available pasture. Moreover, owners of

Yörük women deliver milk to a mobile dairy.

fallow fields now came to charge ever-increasing fees from the nomads. Even summer pastures, most of which were located on village common land, came to be rented. These developments led once again to a large-scale settlement by Yörük households who sought alternative means of livelihood. Some became farmers, other shopkeepers and merchants.

Pastoralism for those who remained nomadic now required larger herds and regular access to substantial amounts of cash and/or credit. Whereas formerly a household could support itself with twenty or thirty mature sheep, soon even 100 came to be regarded as a small herd. The all-cash economy favored households who could invest in veterinary care, in grain feeding, and sometimes in special canvas tents to shelter the herds in winter. Animal husbandry fast became a specialized commercial venture. In this context, it is not surprising that a number of entrepreneurial members of the tribe established mobile dairies. These dairies were moved by trucks to where the animals were grazing in spring and summer pastures. Milk from the sheep was processed daily into white cheese, packaged in tins, and sold to wholesalers in Ankara and Kayseri. Other Yörük became livestock brokers, and bought and sold animals and wool. Although at no time did social and economic differentiation among the still-nomadic households approach that found in most villages, signs of stratification began to emerge among the Yörük. Some families were able to increase their animal holdings and other forms of wealth as others were forced into debt.

Today few of the Yörük remain nomadic in the sense of being tent-dwelling animal herders. Those families who still herd usually maintain a permanent home in town and use hired shepherds who truck their animals to distant pastures. In fact, Yörük herding has undergone a near-complete transformation and resembles in its use of capital what we term ranching in the United States. Rather than being associated with a distinctive style of life and cultural identity, herding is now simply one of a number of investment strategies.

THE BASSERI OF SOUTHWESTERN IRAN

A second case study offers a good insight into the social structure and patterns of leadership of a large tribal grouping in southwest Iran; the study also provides a good sketch of the economics of production and exchange in a pastoral society. Fredrik Barth's monograph on the Basseri continues to be regarded as a classic because of the clarity with which it outlines some of the basic principles that appear to shape pastoral society and economy.[11]

As described by the Norwegian anthropologist Barth, the Basseri numbered some 16,000 people or 2000 to 3000 tents in the mid-1950s. Historically they enjoyed a large measure of political autonomy, a pattern that was not uncommon for the rest of the tribal groups in southwest Iran. Keeping in mind that Barth did his research in the mid-1950s, we will discuss some aspects of Basseri domestic and social life.

The Basseri are a Persian-speaking people in an area where there are tribes speaking Luri, Arabic, and Turkish. The Basseri were the leading group in a tribal confederation called the Khamseh, which until it broke up included Arabic and Turkish tribes. The significance of this large confederation was much diminished since the reign of Reza Shah, who forcibly disarmed and settled nomadic segments in the 1930s. Although the settlement of the Basseri was virtually complete at one point, many families resumed nomadic life at the abdication of Reza Shah in 1941. Today the Basseri are fairly widely dispersed, and the tribe has many settled communities.

The Basseri nomads do not occupy an exclusive territory. As Barth puts it, each of the major tribes of Fars province "owns" the rights to a route, or *il rah,* which it follows each year in its migratory cycle. Rather than owning the land through which they pass and on which they graze their flocks, the Basseri regard the *il rah* schedule as their right-of-way. The *il rah* varies yearly with the availability of passes and roads and with the sequence set by the maturation of the grasses for grazing. The intricate schedule of Basseri yearly migration is coordinated and directed by their paramount chief, the khan.

[11] Fredrik Barth, *Nomads of South Persia: The Basseri Tribe of the Khamseh Confederacy* (Boston: Little Brown, 1961).

The *il rah* of the Basseri takes them from winter pastures in the south on the arid plains or deserts to distant summer pastures at altitudes as high as 6,000 feet. The spring migration northward and upward, as well as the return down in the fall, is marked by frequent breaking of camp. The summer and winter grazing, however, require much less displacement. The Basseri rely primarily on sheep and goats, and like the Yörük of Turkey, keep camels, horses, and donkeys for transport.

Barth considers camp groups to be the primary communities of Basseri nomadic society, analogous to hamlets and villages among peasants. What distinguishes these groups, and we might add all nomadic camp units, is that they persist only as long as members decide to move and work together. Daily consensus is necessary on such vital issues as where to graze and how long to stay in any given area. Barth analyzes the processes by which these collective decisions are made and how the unity of the camp is maintained.

Leadership in the camp is expressed in two ways. One form of leadership is that exercised by headmen, or *katkhoda,* who are formally recognized by the paramount chief of the Basseri. A second type of leadership is informal, one held by elders called "white beards" (*riz safid*), who by consensus are thought of as natural decision makers for the community. The *katkhoda* derives authority in part from the paramount chief, or khan, and in part from the Iranian government, which plays an important role in controlling Iranian tribes. The *katkhoda* relays directives from the khan and from government and reports back to them. Daily management of the camp is in the care of its "white beards" or household heads.

At the time of Barth's study, there were twenty-four named *oulads,* or lineages, the largest of which had 126 tents and the smallest 13. These *oulads* themselves were grouped into larger descent segments called *tira,* of which the Basseri had 15. The *oulad* owns or controls a joint grazing area and a migratory route. Only as a recognized member of one of these lineages does one have access to pastures. The leaders (*katkhodas*) of the *oulads* report directly to the khan and are held responsible to him for the behavior of members of their specific groups. These *katkhodas,* therefore, have considerable authority in their communities.

At the top of the Basseri political hierarchy stands the paramount chief, or khan. For the last three generations the khans have come from one family, constituting something of a ruling dynasty within the tribe. Most members of the khan's family lived sophisticated lives as members of the Iranian national elite. The khan possessed great power and privilege, entertained extravagantly, and at times exercised near-complete authority over his tribal subjects. Until his power was broken by the Iranian government, the ruling khan could impose fines and taxes, conduct court hearings, and order corporal punishment and even death sentences. Closely associated with the khan was a special group known as the *Darbar*. Members of this tribal segment served as the khan's private militia, servants, and companions. Even though the Darbar also owned flocks, tents, and property, they were considered outsiders to the Basseri tribe.

The khan represented the Basseri in their dealings with the Iranian government and in conflicts with sedentary populations. Because he enjoyed great wealth in land as well as in herds, the khan moved with ease among the elite and powerful of the country. He kept houses in Shiraz, the provincial capital, and sent his children to be educated abroad.

In his description of the organization of production among the Basseri, Barth concentrates on the household and on the patterns of wealth accumulation and deployment. Barth surveyed thirty-two tents of one Basseri group and found out that the average household size was 5.7 people, a figure comparable to that for most of rural Iran. Most of these tents were nuclear households. The household is a property-owning unit as well as a social grouping; each depends on its own equipment and animals, and generally on its own labor as well.

The average household herd consisted of about 100 sheep; a typical family could not subsist on fewer than sixty animals, at least in 1958 when the study was carried out. Barth estimates that the average household spent cash for food and other necessities amounting to no more than 40 to 50 percent of the value of its productive capital, or the mature animals. How is this high yield possible? The flocks are comprised primarily of female animals, and the mature ewe annually produces products equivalent to her total market value. Of course, all of this is not profit, as herds have to be replaced, labor hired, and animals may be lost or stolen. Still, this yield is adequate to ensure the average Basseri family a high standard of living relative to many peasant populations.

Barth describes a wealthy 55-year-old man who began his career at age 15 with some twenty sheep probably received as his share of inheritance. His herd grew through natural increase, and soon he was able to sell approximately twenty sheep a year. With the money and with money earned from trading in animal hides, he was eventually able to buy land and settle in a village. By then his herd had grown into several hundred sheep, and he had a number of shepherds to help. This pattern of wealth accumulation, investment in land, and subsequent settlement is not unusual. Nomadic families whose herds prosper normally try to reduce their risk of sudden loss and impoverishment by investing in other resources. Many households, Barth notes, may acquire wealth and settle, only to resume nomadism several times throughout their careers. On the other hand, poor families whose herds fail settle out of necessity and hope that with time they will be able to save enough money from tenant farming and wage labor to resume pastoralism. The latter holds at least the promise of high returns. The fact that both the rich and the poor regularly leave the nomadic segment of the Basseri limits extreme differentiation in wealth among the pastoralists.

One limit to the accumulation of wealth in animals, Barth notes, is that animals do best in the care of their owners. Reliance on hired shepherds means higher rates of loss due to poor care or theft, and encourages the investing of surplus wealth in land. The relatively homogeneity in wealth reported among the Basseri is by no means common to other groups. Daniel Bradburd and Lois Beck, working separately among the Komachi and Qashqa'i tribes, report great disparities in wealth and social class. In fact, in both societies a class of poor men are employed as shepherds by the wealthy herd owners.[12]

Although we have no recent reports on the current status of the Basseri, Beck has visited the Qashqa'i since the Iranian Revolution. On the eve of the Revolution, the larger tribal structure had already been greatly weakened by the state, and smaller tribal segments assumed greater responsibility for their well-being and security. Land-reform measures, changing patterns of land use, and the availability of wage labor all contributed to restricting pastoralism and the importance of the tribe. Beck suggests that the chaos created by the Revolution of 1978 has led to a resurgence of pastoralism as well as tribalism. In the absence of government agents, tribal leaders are reassuming some of their traditional functions.

Al-MURRA OF SAUDI ARABIA

Al-Murra of the Empty Quarter of Saudi Arabia come closest to our notions of the classic desert nomad, the Bedouin.[13] Indeed the Murra refer to themselves as *Bedu al-Bedu,* or literally, "the nomads of the nomads." They are among the last of the great camel-herding nomads of the Arabian Peninsula.

In 1968 Donald Cole, an American anthropologist, began a study that lasted two years and that provides us with some of the best data on the contemporary Bedouin and how they fit into the society of oil-rich

[12] Lois Beck, "Herd Owners and Herd Shepherds: The Qashqai'i of Iran," and Daniel Bradburd, "Size and Success: Komachi Adaptation to a Changing Iran," in *Modern Iran: The Dialectics of Continuity and Change,* eds. Michael Bonine and Nikki Keddi (Albany: State University of New York Press, 1981), pp. 123–137.

[13] Donald Cole, *Nomads of the Nomads.*

Saudi Arabia. Cole has since returned many times to that country and continues to report on the transformations there.

Al-Murra, a tribe of some 15,000 people, make up part of the estimated 20 percent of Saudi Arabia's total population that was still nomadic in 1968. They live in one of the most forbidding desert regions of the world. The group studied by Cole, Al Azab, keep camels, whereas other sections of Al-Murra herd sheep and goats as well. The camel is much more than simply an animal kept for production. In Cole's words, it is "an abiding passion of the tribesman," who tell stories and make up songs and poems about their favorite camels. The camels of the tribe are famous for their milk, which is an important dietary item, along with dates, rice, and bread. Apart from milk, all these are purchased in market towns, especially at the oasis of Hofuf. Before the development and expansion of the oil economy, Al-Murra obtained most of their food needs through trade or tribute. Today, as members of the Saudi Reserve National Guard, Al-Murra tribesmen draw a salary from the government. Another source of cash income is from the wages received as laborers in the oil fields.

Al-Murrah claim a territory, *dira,* which extends in an arc across the west and central borders of the Rub al-Khali. Several oases fall within the *dira,* including those of Nejran and al-Hasa. They also claim exclusive rights to a series of wells in the central part of this territory. The outer margins of the *dira* are not clearly delineated and overlap those of other tribes.

Nomadic pastoralism in Arabia relies on the yearly cycle of rainfall and on underground sources of water which supply the infrequent wells and determine what pastures can be used. During the summer months the camels are virtually untended while their owners camp near the wells. The far-ranging grazing animals come back on their own every four days for water, and at this time they are also milked. In an average year the Al Azab section of the Murra traverse a distance of approximately 1200 miles, which comprises their annual migratory cycle. Of the two major migrations, spring and fall, the spring is the time of greatest concentration of tents and hence social interaction. Feasting and celebration of marriages take place then, and the great *sheikhs,* or royal princes of Saudi Arabia, visit the camp and are feted.

A camp group among the Murra is made up of a number of households, or *beits* (singular: *beit,* house or tent). Each *beit* is thought of as possessing or sharing a unique and private domestic space, called the *dar.* Sometimes two or more *beits* will share a single *dar,* as for example when two married brothers regularly camp and move together. The major meal of the day is eaten in common by members of the *dar,* although herding and animal care are handled by the individual households.

Each tent tends to be an autonomous social and economic grouping, free to move and to form associations with any other tent. These associations, although usually based on kinship, may also be formed through ties of friendship or convenience. Sometimes tents attach themselves to others to avail themselves of the protection that neighbors customarily extend to each other.

The fact that the *dira,* or tribal range, its wells, and its other resources are associated with distinct lineages of the Murra means that most camp groups are composed of close agnatic kinspeople. Al-Murra lineage, or *fakhd,* is made up of a number of people who claim common patrilineal descent and who share certain rights and responsibilities. The *fakhd* "owns" the wells and the *dira* range and is further responsible for collecting or paying blood money in the case of a murder. The herds of *fakhd* members are marked with the same brand, a geometric sign called *wasm,* which is burned into the flanks of the camel. In former days the *fakhd* was an important organization for military action, including defense as well as raiding.

A number of lineages group themselves into larger segments called *gabila,* or clans. There are seven *gabilas* within the Murra tribe. Today these named groups play little role in Al-Murra life, as they own no resources in common and share no real responsibility. They serve primarily as reference groups identifying related lineages. Historically the *gabila* was important in political life. Each was headed by a leader, called *emir* to reflect his special military status. In times of large-scale warfare the *gabila* united behind its emir, and on occasion the entire Murra grouped under the military leadership of one of the emirs.

The entire tribe, or the people of Al-Murra taken together, share a strong sense of common heritage going beyond that of descent. They speak one dialect, dress in a distinctive style, and have their own manner of furnishing their tents. Marriage is almost always within the tribal group; and they put it, *"Al-Murra, kuluna wahid, biyutuna wahid, kuluna ikhwan"*—that is, "People of the Murra are all one, our tents are one, we are all brothers."

Until the ascendancy of the Saud family in the late nineteenth century, the Murra tribe was an effective military group capable of conducting raids and warfare. Today it is one reserve unit within the Saudi Arabian National Guard. In this century, the paramount leaders of the tribe, or emirs, have come from one clan. The present emir lives in a black tent (at least he did in 1970), but rarely participates in the annual migration. His complex of tents is situated not far from the town where he also keeps a house. Tribespeople can find him easily as they come and go from the town. Like the Basseri khan, he entrusts his herds to the care of his fellow tribespeople.

The emir serves as commander of the Murra Reserve Guard Unit; he distributes government salaries to the tribespeople. He also acts as arbitrator in disputes within the tribe and attempts to deal with problems such as those arising from car accidents, assaults, and murder. He intervenes in every case in which a tribesperson is held by the police, and meets with high government officials or members of the Royal Court to resolve such cases.

Tent life resembles that of the Yörük and the Basseri. The tent is the domain of the women, and as men increasingly leave for work in towns, women form the resident element of the household. Cole writes that the Al-Murra express a strong preference for close kin marriage, the preferred one being between a man and his *bint 'amm*, or paternal uncle's daughter. Cole provides no statistics, however, on the actual occurrence of this form of marriage. Marriages are decided upon by the fathers of the couple, but negotiation may be initiated by the mothers. It is the men who have to agree upon the amount of the bride wealth, or *mahr*, which varies with the degree of kinship involved and the social status of the families.

The interior of a tent is separated into men's and women's sections. Women move freely in the men's section when guests are not present and can visit markets in the oasis towns. The division of household labor parallels that found elsewhere in the Middle East, with the exception that Al-Murra men do the milking.

GENERAL OVERVIEW

In our discussion of nomadic pastoralism and in our three case studies, we have touched on many aspects of nomadic life, as well as on the economic and political organization of pastoral communities. Throughout we have tried to avoid stereotypes and to introduce the variability and diversity that exist among nomadic peoples. We have stressed that the nomadic peoples of the Middle East are now, and historically have been, part of a larger social and political world. Economically they were always integrated in larger systems of exchange. Politically they have played a historic role in the formation and overthrow of ruling dynasties. Any attempt to see either tribalism or nomadic pastoralism as a vestige of the past is therefore misleading.

The political nature of tribalism, which is a topic quite separate from nomadic pastoralism, is taken up elsewhere. What is important to emphasize here is that tribalism among nomadic peoples, even for those like the Yörük and Basseri who have lost any form of political autonomy, remains an important means of expressing individual and group identity in many ways analogous to ethnicity.

We have also emphasized that pastoralists today are very much twentieth-century people, facing the full range of problems and opportunities of modern society. To this extent their ways of making a living and their social and political organization have changed rapidly in the last few decades. Pastoralists, like all other rural people, are increasingly dependent on mechanized transport, distant sources of industrial goods, and rapid communications by post and telephone for market and other information. Different people have, of course, responded in different ways. The Yörük of Turkey have become small-scale ranchers, behaving in ways that directly resemble the activities of ranchers everywhere. They phone livestock brokers to get market quotes, arrange truck transport of animals, go to the banks for credit, and even sell futures in immature stock. Al-Murra, on the other hand, living in an oil-rich state, rely on government salaries and have probably decreased their reliance on their animals as an economic investment. The Bedouins of the Negev appear to regard their animal herds as a hedge against inflation and rely primarily on wage labor or farming.[14]

Is there a future for nomadic pastoralism in the Middle East? If this is taken to mean the age-old movement of flocks and tents united in a distinctive style of life, the answer is probably "no." Marginal fields once jointly exploited by both farmers and herders are now put to intensive cultivation. Irrigation, modern seeds, machinery, and fertilizers make this possible. Many of the routes to mountain pastures in the Zagros, Taurus, and other ranges are now interdicted by such obstacles as national borders, zones of intensive cultivation, or political restrictions on movement. For most regions, the number of people who can camp together in any season is reduced because of the larger number of animals needed to sustain a household now, as well as diminished pasturage. This, of course, affects the nature of nomadism as a special way of life. Also, for most groups mechanized transport means that families no longer experience large-scale migrations together and the sense of shared experience that this generates. Even with its hardships, most Yörük seem to feel that the easy sociability of camp life among tents of close friends and relatives was something to be prized. Animals continue to support people, but they are coming less and less to support distinctive communities or local cultures.[15]

[14] Emanuel Marx, "Economic Changes Among Pastoral Nomads in the Middle East," paper presented at the Conference of the Commission on Nomadic Peoples, International Union of Anthropological and Ethnological Sciences, London, 1978.

[15] For a useful collection of studies concerning nomadic settlement, its causes, and its consequences, see Philip Salzman and E. Sadala, eds., *When Nomads Settle: Processes of Sedentarization and Response* (New York: Bergin Press, distributed by Praeger, 1980). See also Cynthia Nelson, ed., *The Desert and the Sown: Nomads in the Wider Society* (Berkeley: Institute of International Studies, University of California, Research Series No. 21, 1976).

CHAPTER SIX: AGRICULTURE AND THE CHANGING VILLAGE

The Middle East is one of the areas of the world in which agriculture developed earliest—some 10,000 years ago.[1] Elements of this long agricultural tradition persist even today. Historically the majority of the inhabitants were rural dwellers and depended on agriculture for their livelihood. Today this situation is changing. In 1938 almost 75 percent of the inhabitants of Egypt were village residents. In 1976 an estimated 56 percent were rural, and the figure continues to change rapidly as a result of the massive influx of people to cities. This trend is replicated throughout the area, and very shortly, if not already, the bulk of the people in the region will reside in urban communities. Despite this, agriculture remains important, and village life and society continue to form key elements in the Middle Eastern social landscape.

In 1938 a Jesuit priest, Father Ayrout, wrote what is still considered a classic book on the Egyptian fellahin, or peasants. In his words, the fellahin, members of one of the world's oldest civilizations, can justifiably be called The People of Egypt. He describes them as people whose masters, religion, and language have changed with time, but whose way of life remained very much the same: "as tranquil and stable

[1] See Chapter 1.

as the bottom of a deep sea whose surface waves are lashed with storms."[2] Everywhere in Egypt, writes Ayrout, one sees the same agricultural tools—the plough, water wheel, winnowing fork, and sickle—used since Pharonic times. Similarly, in his view, the social life of the fellah shows remarkable persistence, dominated as it is by the land-owning master on the one hand, and by the relentless routine of irrigation agriculture on the other. The Egyptian peasants, as described by Ayrout, emerge as timeless and yet without a history. Tied to changeless agricultural routine and at the mercy of exploitive masters, they found refuge in a self-contained and conservative cultural universe, defined largely by family and faith.

Although Ayrout stressed the uniformities and continuities in peasant life, considerable regional and social variation as well as historical change are equally true of Egyptian rural life. Jacques Berque, a French sociologist, provides a detailed account of the social history of one Egyptian village in the twentieth century which puts things in a different light.[3] He documents its changing political structure and how its people regularly altered their strategies of production and ways of life as conditions demanded. Far from a passive and unchanging picture of the "eternal fellah," Berque gives us a more realistic portrait of the inhabitants of the village of Sirs al-Layyan as families and clans who contend for social status and acquire and lose political power in a changing economic environment. This picture of rural life in Egypt more closely resembles that of peasant communities in other parts of the world.

Lucy Saunders, an anthropologist who has worked in an Egyptian Delta village over the past 10 years, has described the profound social and economic changes which are taking place in the area. Government programs, services, and regulations are increasingly relevant to daily life in the village. Even the landless tenant now has legal rights to the land he or she rents, which dramatically curbs the influence of the old land-owning families. Today a large number of village youths have completed college, and increasing numbers of peasants find employment in urban centers and nearby brickmaking factories. Large-scale poultry production, newly introduced, has made a few families wealthy, while others, as before, remain poor. In short, the village Saunders studied is changing every bit as much as is the rest of Egypt. Everywhere in the Middle East, village life is being affected by the political and social transformations underway, and rarely can we speak of village life as unfolding in self-contained isolation.[4]

[2] Henry Habib Ayrout, *The Egyptian Peasant* (Boston: Beacon Press, 1963; originally published 1938), p. 1.

[3] Jacques Berque, *The Social History of an Egyptian Village in the Twentieth Century* (Paris: Mouton, 1957).

[4] Lucy Saunders, personal communication to the authors.

In the course of this chapter, we touch on many of the changing characteristics of village society and economy. These include a near-complete shift to capital-intensive, privately organized farming, the increasing integration of village economies into national markets, and changes in land tenure and access to land. These processes have led to the narrowing of certain aspects of the social and cultural gap that has historically separated urban and rural societies in the Middle East. They have also led to increased socioeconomic differentiation within rural society itself, all of which have profound implications for the region as a whole.

LAND USE AND RURAL SETTLEMENTS

Agriculture and patterns of rural settlement are strongly conditioned by two environmental factors: availability of water and topography. The distribution and amount of water determine the possibilities of agriculture, and hence the distribution of peoples. Regarding topography, the question is simply whether sufficient arable land exists in plots large enough to warrant human investment in capital and labor. For example, in northeastern Iran, in an area formerly occupied by Turkmen nomads, many new villages have been established in regions where few permanent settlements existed until recent times. Quite apart from the problem of political security, agriculture was limited by the fact that traditional technology did not make it worthwhile to cultivate the extensive tracts of open land that received adequate rainfall only on an irregular basis. Well-watered areas in the region were largely limited to valley bottoms, but these were usually insufficient in size to sustain a specialized agricultural population. Land drainage and modern technology have vastly increased the scope of agriculture. Prior to this, farming was an endeavor secondary to pastoralism; such limits typified many areas recently cultivated in the Middle East.

Even though there might have been a high yield in some areas in certain years, the extreme unreliability of water placed limits on the size of the population and its distribution. Where water was regularly available and in sufficient quantity, as in the valleys of the Nile and the Tigris-Euphrates and in some coastal regions, intensive agrarian economies emerged. Again the prime example is that of Egypt. The soil of the Nile Valley is of unusual fertility; after each annual flooding a new load of silt carried from the head waters of the Nile in Ethiopia and Uganda is (or was, prior to the new High Dam in Aswan) deposited to replenish the fields of Egypt. Productivity is extremely high, allowing multiple cropping and sustaining a dense population clustered around the river banks and canals.

Until the construction of the large-scale dams to control and distribute water on a year-long basis, the Egytian peasants depended on the annual floods in a regimen dating back to the time of the Pharaohs. The old regimen consisted of dividing the fields up into shallow basins to capture the flood waters by building earth embankments around them. One crop a year was planted, usually a winter crop such as wheat, barley, and fodder. Summer crops like cotton, rice, maize, and millet were not well suited to basin irrigation and only spread following the introduction of perennial irrigation. Cotton, the premier export crop of Egypt today, dates from the development of perennial agriculture in the midnineteenth century. The Aswan High Dam, completed in 1970, has extended year-long cultivation even longer.

A situation similar to that of the Nile Valley exists in patches or zones along the meandering courses of the Tigris and Euphrates rivers, which discharge their waters in April and May. Iraq's irrigation problems, however, are rather special. Unlike Egypt, many crops are planted prior to peak flood. The late flooding of Iraq's two major rivers mandates the impounding of the waters within their banks in order to prevent large-scale destruction of the maturing crops. Moreover, poor drainage of stagnant floodwaters exacerbates the problem of soil salinity. One of the major development projects in the Middle East today is one designed to desalinate vast tracts of soil in southern Iraq that have been contaminated by years of inadequate drainage. The problem of maintaining soil quality continues to plague some of the most productive agricultural areas. All too often irrigation agriculture has made it too easy for the farmers to overirrigate.

Along the southern Turkish coast, there are well-watered, densely populated agricultural areas. For example, on the broad plains around Adana, Turkey's fourth largest city, we see a dense network of villages all closely integrated into a national market of cotton, tobacco, and tree crops. In coastal Turkey, the Fertile Crescent, and elsewhere, such dense settlements often occur in the same plains and valleys that were important to the city-states and empires of antiquity.[5]

Although this pattern of land use may appear as one of the examples of continuity in the area, the facts are more complex. These same regions of Turkey as well as those along the Tigris-Euphrates valley that are so densely populated today were virtually lost to agriculture in recent history. In effect, many areas of the Middle East are being resettled

[5] Mustafa Soysal, *Die Siedlungs und Landschaftsentwickelung der Cukurova*, Erlanger Geographische Arbiten, Sonderbank 4 (Fränk-ishen Geographischen Gesellschaft. Erlanger, 1976); and Mubeccel Kiray, "Social Change in Cukurova: A Comparison of Four Villages," in *Turkey: Geographic and Social Perspectives*, eds. Peter Benedict, Erol Tümertekin, and Fatma Mansur (Leiden: E. J. Brill, 1974), pp. 19–47.

today. Cases in point include the central plateau of Turkey, the Syrian steppes, and parts of north and south-central Iran.[6]

High rates of population growth are obviously related to this expansion of rural settlement and intensification of land use. However, population pressure is not the only factor. Political and economic considerations are equally important. They include the ability of governments and settled communities to maintain security and lines of communication. Even more important is the fact that the availability of urban-controlled credit and market facilities has encouraged rural investment and the transition to modern farming. This is not an entirely new phenomenon. In the past as well, rural settlement and agricultural production were closely tied to urban investment in such things as irrigation facilities, markets, and, of course, roads.

One study that illustrates how these factors are interrelated is Paul English's work on the Kirman region in central Iran.[7] English, a geographer, focuses on the patterns of interdependence among the villages of the Kirman basin and the relationship of these villages to the city of Kirman. He suggests that the dominance of Kirman City is the key to understanding the settlement history and economy of the region. Village size and complexity decrease with distance from the regional capital of Kirman. The settlement history of the region is traced by English back to the installation of a Sassanian garrison center in the area in around A.D. 240. This was soon followed by the establishment and spread of villages as a result of increased security and the deployment of urban capital by the land-owning elite of the city. This capital was used to build the elaborate underground irrigation networks (*qanats*) and to dig the wells that provided the water on which village agricultural life depended. Moreover, urban capital was also used to develop a carpet industry based in the villages around Kirman. English's thesis is that urban centers both create village settlements and help to keep villages relatively underdeveloped by the subsequent systematic extraction of surplus and products.[8] Recent land-reform measures in many countries give more political and economic powers to local rural communities, but do not seem to be sufficient to reverse the historical pattern of urban domination. English's conclusion that Middle Eastern village life and settlement

[6] Daniel G. Bates, "The Middle Eastern Village in Regional Perspective," in *Village Viability in Contemporary Society*, ed. Priscilla Copeland Reining and Barbara Lenkerd, AAAS Selected Symposium 34 (Boulder: Westview Press, 1979) pp. 161–83.

[7] Paul Ward English, *City and Village in Iran: Settlement and Economy in the Kirman Basin* (Madison: University of Wisconsin Press, 1966).

[8] For another study of a city and its hinterland in Iran, see V. F. Costello, *Kashan: A City and Region of Iran*, Centre for Middle Eastern and Islamic Studies, University of Durham (Durham, 1976).

patterns are predominately shaped by forces emanating from urban centers appears to hold true.

A Turkish geographer, Mustafa Soysal, gives an account of the development of villages and agriculture around the city of Adana.[9] Soysal demonstrates that the extent to which a village specializes in such crops as cotton is closely related to urban-based control of the land and investment in mechanization. Such communities tend to be politically dominated by the families who, while living in the city, own and control village lands. Marginally located communities, far from major roads and in less fertile areas, though apt to be poor, maintain both local autonomy and a greater degree of economic self-reliance. They also tend to be smaller in size. In short, there is no such thing as a "typical village," but rather each region has its own constellation of differentiated villages.

Regional approaches to village studies can also facilitate an understanding of rural demographic processes including migration. In a study of the Kashan region, for example, Costello found significantly higher rates of outmigration from upland hamlets, where poor soil and lack of capital inhibits agricultural intensification.[10] Although this may be true for the Kashan region, in other areas such as that of Adana, Turkey, mentioned earlier, some of the highest rates of outmigration emanate from the larger and more agriculturally productive communities. Here investment in agriculture and consolidation of holdings into larger farms have forced farmers off the land, some to become migrant laborers, others to move into cities.[11] The point to remember is that even though migration or movement between villages and from villages to cities is a regular phenomenon, it will vary among villages depending in part on the location of settlements within a regional system.

LAND TENURE

Agricultural production, as we noted in the preceding discussion, depends on a number of key factors, among which are land, water, labor, and technology. Although all are important, because of its enduring quality and limited availability, arable land is the most regulated by law and custom. How people get access to the land they use and how they derive profit from the land they hold rights to constitute the system of land tenure. The systems found in Middle Eastern countries, while varying locally, all traditionally have a common base in Islamic law, as

[9] Soysal, *Die Siedlungs und Land.*

[10] Costello, *Kashan.*

[11] Soysal, *Die Siedlungs und Land.*

modified to meet local conditions and customary usage. More recently, the codification of land-tenure practices in many countries has incorporated a number of European legal concepts, of which the most important is perhaps the registration of individual title.

There are four major categories of land rights found throughout Arabic- and Turkish-speaking areas of the Middle East. Persian categories, while essentially similar, have different terms. The Arabic terms are: *mulk, miri, waqf,* and *musha'a.* We now briefly describe each and give examples of how these concepts operated in practice.

The category of *mulk* refers to freehold or private ownership. This is similar to American notions of private property and entails full rights of use and disposal through sale or inheritance. The concept of *mulk* extends to all kinds of property. In the Middle East today private or *mulk* ownership of land is the most prevalent form of tenure. This was not the case prior to the nineteenth century, when current evidence suggests that the bulk of agricultural land was ultimately the domain of the state or the ruler.[12]

State or Crown lands are generally known as *miri,* although this category subsumes a wide variety of actual practices. For example, within the Ottoman Empire two major forms of *miri* land tenure were common: the *timar* and the *iltizam.* On richer lands, the *timar* was dominant until the eighteenth century. In this system, agricultural estates were granted to individuals in compensation for services rendered to the ruling dynasty. In return for the land, the grantee had to raise troops for the sultan's army upon demand. In theory, the land could not be passed on as an inheritance, but reverted to the state once the grantee died. In this way *timar* differs significantly from European feudal practice. The rule against inheritance was often violated, but where effective control remained with the sultan, this practice blocked the development of a stable landed aristocracy, which was a prominant feature in European history. Of course, inheritance patterns also limited the development of a landed aristocracy as the rule of partible inheritance broke up large estates.

The second major form of *miri* land administration was known as *iltizam,* or tax farming. In this system land theoretically owned by the state was turned over to tax farmers, a practice that was particularly prevalent in the Ottoman province of Egypt. The tax farmer, usually an

[12] Research into the economic history of the Middle East and especially land-tenure regimes is still lacking on any large or comparative scale. The best-researched area from this perspective is Egypt. See Gabriel Baer, *Studies in the Social History of Modern Egypt* (Chicago: University of Chicago Press, 1969). A book of readings edited by Charles Issawi, *The Economic History of the Middle East, 1800–1914* (Chicago: University of Chicago Press, 1966; reprinted in paper in 1975) is an invaluable source for this crucial period.

individual in favor at the court, paid a fixed price to the sultan for the right to tax the peasants of a particular tract of land. Both *iltizam* and *timar* represented forms of absentee control over land, its resources, and the peasantry who labor on it.

The *timar* system was abolished at the beginning of the nineteenth century, and so was the *iltizam* soon thereafter. The Ottoman government moved into direct collection of taxes from the peasants and attempted to encourage private ownership of land. This was done not so much to benefit the peasant, as to ensure more direct control of the land, higher productivity, and increased revenues. This process can perhaps be seen most clearly in the case of Egypt where, beginning in 1811, the ruler Muhammad Ali embarked on a series of agrarian reforms that culminated in the Land Law of 1858. Similar reforms were instituted elsewhere in the Middle East at about the same time. These reforms were all designed to limit the power of the elite and to promote agricultural production through private ownership of land. As Kenneth Cuno shows, these reforms did not arise in eighteenth-century Egypt in isolation from other forces already changing agrarian society. Even prior to initial attempts at legal land reform, private ownership of a de facto sort—that is, not based on governmental decree—was gaining ground, stimulated by rising prices of foods and food export to Europe. Thus, in many instances legal reform was as much a consequence as a cause of shifting patterns of tenure.[13]

During this period, and at least related to this transition to private ownership, Egypt experienced a very rapid transition to cash crops and an expansion of its network of irrigation.[14] This allowed those with capital to purchase lands and rapidly gave rise to a new basis for social differentiation, that of land ownership. The newly emergent classes may be roughly grouped into the following categories: urban-dwelling landlords who administered their estates through supervisors or agents, relatively small-based land owners, landless tenants or sharecroppers, and finally the landless wage laborers. Concurrent with these developments, the shift from subsistence agriculture to cash crops meant that the farmer was increasingly tied into the world market system. It is ironic that many subsequent attempts at land reform in Egypt, as elsewhere in the Middle East, have been directed at remedying the social inequities arising from this first effort of land reform.

Besides the *miri* and *mulk*, a third category of land tenure is the *waqf*, which refers to an Islamic institution whereby a property is designated as a self-perpetuating trust whose income is then assigned to

[13] Kenneth Cuno, "The Origins of Private Ownership of Land in Egypt: A Reappraisal," *International Journal of Middle Eastern Studies*, (1980), 245–75.

[14] Baer, *Studies in the Social History of Modern Egypt*.

some end. The most usual form of *waqf* is the endowment of a charitable endeavor. Income from a piece of land, a shop, or some other form of property is committed to the support of a mosque, shrine, hospital, or school. For example, until recently, most of the major mosques of Istanbul were maintained by *waqf* lands in Syria, Palestine, and Egypt. The lot of the peasants who lived on these lands differed very little from that of tenant farmers anywhere else.

Another type of *waqf* is private. Individuals could designate up to one-third of their property as a trust for their descendants. This was aimed at preserving family estates by circumventing fragmentation through inheritance and alienation through sale. The net effect of the *waqf* system on land tenure and land use was to create large estates administered by bureaucrats who had no direct stake in their productivity or maintenance, not to mention reinvestment and capital improvement.

The economic inefficiency and social abuse associated with the *waqf* encouraged modern governments in the area to attempt to abolish it or curb it. Land reform laws in Egypt, Iraq, Syria, Turkey, and Iran, to name a few countries, moved forcibly to break up *waqf* estates, especially the private ones. In Egypt one unfortunate by-product of this has been the increasing neglect and deterioration of the country's great Islamic monuments. No longer supported by their traditional *waqf* revenues, they now have to depend on an overburdened public treasury. In Iran the late Shah's attempt to break up the vast *waqf* holdings met with strenuous and ultimately successful opposition from the religious establishment.[15] In general, though, this form of rural tenure is in decline.

The last and least common form of land tenure is the *musha'a,* or commons. *Musha'a* designates the common tenure and periodic redistribution of village lands. Generally speaking, *musha'a* did not exist in isolation from other forms of land tenure but rather in conjunction with them. For example, agricultural lands, especially orchards and irrigated fields, may be held as *mulk,* or private property, while the adjacent pastures are held in common, or as *musha'a.*

Musha'a as a significant form of agricultural land tenure is, or rather was, found primarily in the Levant, often in areas marginal to rainfall agriculture and occupied by recently settled tribal communities. Few such communities have maintained this form of tenure in the post-World War II period; one of the last areas where it was prevalent was

[15] For additional readings on land reform in Iran, see Ann K. S. Lambton, *Landlord and Peasant in Persia: A Study of Land Tenure and Land Revenue Administration* (London: Oxford University Press, 1953); Ann K. S. Lambton, *The Persian Land Reform* (London: Oxford University Press, 1962); and Nikki Keddie, "The Iranian Village Before and After Land Reform," *Journal of Contemporary History,* 3 (July 1968).

Palestine. Where most of the village lands were held as *musha'a*, it was usual to periodically redistribute fields among families according to familial need as well as to equalize access to better plots.[16] Today *musha'a* is usually limited to unproductive or grazing land.

TENANCY AND ACCESS TO LAND

So far we have discussed landholding categories without saying much about how people actually get access to the land which supports them. One has to distinguish carefully here between those who own the land and those who actually cultivate it. Historically the majority of Middle Eastern peasants probably belonged to the second category—that is, they cultivated land they did not own or control, and their access to land was through a number of tenancy arrangements.

Different types of tenancies have long characterized Middle Eastern agricultural relationships and persist even today, despite the recent land-reform efforts. Historically the most prevalent form of tenancy is sharecropping, whereby the cultivator provides labor and sometimes animal traction and tools, in return for a percentage of the crop that he has raised on land belonging to someone else. The size of the share varies according to what each party provides. Sharecropping was so common a practice that the term for peasant throughout most of the area was synonymous with that of the most common form of sharecropping, the *khemmas*, or the one-fifth. In this form of sharecropping, allocation of shares was based on the five components of production: land, water, animal traction, seed, and labor. A peasant supplying only his labor would receive one-fifth of the harvest; if he provided animal power as well, he would receive two-fifths, and so on. Direct rental of land for a fixed fee was rather rare until very recently. Even on large estates, land was usually let out as small scattered plots in return for a share of the harvest, a practice that still persists in many villages.[17]

Sharecropping has some advantages for both owner and tenant under circumstances where yields are highly variable from one year to

[16] For a detailed study of a village still holding to musha'a tenure, see Barbara Aswad, *Property Control and Social Strategies: Settlers on a Middle Eastern Plain*, University of Michigan Anthropological Papers No. 44 (Ann Arbor: University of Michigan, 1971).

[17] Nicholas S. Hopkins, *Animal Husbandry and the Household Economy in Two Egyptian Villages*, a report of the Rural Sociology Segment of the Project on Improved Utilization of Feed Resources of the Livestock Sector, mimeo, Social Research Center, American University, Cairo, May 1980; Mahmoud Abdel-Fadel, *Development, Income Distribution and Social Change in Rural Egypt, 1952–70: A Study in the Political Economy of Agrarian Transition* (Cambridge, Eng. and New York: Cambridge University Press, 1975), p. 143.

the next. Under this arrangement the cultivator need not provide or maintain much equipment, draft animals and the like, and should seed supplies be difficult to obtain locally the cultivator could rely on the landowner to provide them. Moreover, the cost to the cultivator was automatically adjusted in terms of the annual yield. In disastrously bad years, the landlord might forego his share in order to maintain the peasant-laborer and might even advance the peasant-laborer food supplies.

Relationships between the peasant and the landlord traditionally took the form of generalized patron-client ties. In this relationship, the landlord assumed wide responsibility for assisting and protecting the tenants. He may have interceded on their behalf with government officials, or he may have assisted them in medical and other crises. However, one must not romanticize this feudallike relationship. In his novel about Turkish rural life, *Memed, My Hawk,* Yashar Kemal describes the misery and suffering that could result from this essentially exploitive relationship.[18] In this portrayal the peasant family lives in fear of dispossession, has little to show for its endless efforts, and is subject to arbitrary abuse by the landlord or his retainers. The landlords, or *aghas* as they are called in Turkey, may exercise near-total control over the lives of those who work their land.

Throughout the area where sharecropping is practiced in the absence of government regulation and supervision, it tends to be associated with an exploitive pattern that is manifested in the repeated cycle of indebtedness and poverty of the peasant. Nowhere, perhaps, was this pattern more evident than in Iran prior to the mid-1950s, when the majority of the villages were owned by a relatively small number of absentee landlords.[19] Even as reforms in most countries moved to break up these large estates, in many instances, the introduction of modern agricultural techniques worsened the plight of many farmers. Labor, the one commodity the landless villager can supply, had become less important in the modern context, and many small holders, even when granted land by governments, cannot effectively compete with those who have access to machinery.

So far our discussion of Middle Eastern agriculture and rural life has concentrated on placing the village and agrarian economy within its larger historical and economic context. Before we take up the question of the nature of rural transformation, we shall briefly consider some aspects of the social and political organization of rural society in order

[18] Yashar Kemal, *Memed, My Hawk,* trans. Edouard Roditi (New York: Pantheon Books, 1961).

[19] See Ann K. S. Lambton, *Landlord and Peasant in Persia* and *The Persian Land Reform;* and Nikki Keddie, "The Iranian Village Before and After Land Reform."

to convey some idea of the ways in which rural people organize their domestic and political lives.

VILLAGE POLITICAL ORGANIZATION

Formal leadership in the village is typically vested in the office of the village headman, called *mukhtar* in most Arab countries and in Turkey, *oumda* in Egypt, and *katkhoda* in Iran; these leaders are discussed again in Chapter 10. The *mukhtar* or *katkhoda* is generally elected by the heads of households subject to confirmation by the central government or, in some cases, he may be directly appointed by the government. The office itself is created by the state and reflects efforts to control rural populations. Thus the *mukhtar* represents the government to the villagers and only secondarily serves to bring the village interests to the attention of the central government.

On any weekday in any major Turkish town, for example, the corridors of the district governor's *kaymakamluk*, or headquarters, are filled with *mukhtars* from the surrounding villages. They come to respond to government decrees, to file reports, or to register births, deaths, and marriages. They may be accompanying conscripts as they report to military induction centers, or they may be simply responding to the complaints of their village school teachers, for whose well-being they are responsible. They bring village complaints to the attention of the officials and try to gain support for themselves and their village. If the governor decides that the village streets are impassible, for example, he may require the *mukhtar* to organize a work party, or if necessary, to collect money to repair the village schoolhouse. Should disputes arise among the villagers, the *mukhtar* is expected to mediate, and should open conflict erupt, to call on the assistance of the national gendarmerie. The *mukhtar* is the first and last point of call for foreigners in the village, and he must report to the governor on their presence.

The real status and power of the village headman varies greatly depending on the actual influence of the government in the countryside and the presence of competing local sources of power, such as big landlords or tribal leaders. In villages remote from urban centers, the headman may represent the government's interest less than those of the local power structure. That is, he may exemplify and exercise power in his own right or on behalf of local tribal leaders or landlords. In other communities he may represent little more than a low-level government bureaucrat, wielding little power on his own and simply acting as a messenger for the national authorities. As Paul Stirling, an English anthropologist, has observed for a village in central Turkey, "The head-

Villagers gathered in front of the district governor's office in Eastern Turkey.

man is not as much at the top of the village as at the bottom of the official state hierarchy."[20] Being at the bottom of the state hierarchy, however, need not always mean a lack of power. Although Turkish *mukhtars* tend to be relatively limited in the scope and effectiveness of their activities, those of Ba'athist Iraq and postrevolutionary Egypt, for example, have been delegated considerable authority and power. They derive this, in part, from their membership in the ruling party.

The political dynamics of Middle Eastern village life are obviously much too complex and varied to allow for easy generalizations. In an effort to clarify some of the political patterns, Robert Fernea suggests a system of classification that derives from the differences in the internal organization of villages as well as their relation to the national government.[21] We cannot give examples of villages illustrating all of the many

[20] Paul Stirling, *Turkish Village* (New York: John Wiley, 1965), p. 65.

[21] Robert Fernea, "Gaps in the Ethnographic Literature on the Middle Eastern Village: A Classificatory Exploration," in *Rural Politics and Social Change in the Middle East,* ed. Richard Antoun and Iliya Harik, (Bloomington: Indiana University Press, 1972).

types Fernea suggests, but we can emphasize the significance of one distinction, that of tribal versus nontribal.

TRIBALLY ORGANIZED VILLAGES

In tribally organized villages, access to power, positions of leadership, and available resources are all associated to one degree or another with membership in a named descent group. Fernea illustrates this with the example of the el-Chebayish, a tribal community in southern Iraq studied in the mid-1950s by the Iraqi anthropologist Shaker Salim.[22] Here we rely directly on Salim's fine report.

During the long period of Ottoman rule, the el-Chebayish, who lived in the marsh area midway between the Tigris and the Euphrates, were relatively independent. There was no government agent in the village until 1893, and the authority of the sheikh or tribal leader was paramount. The inhabitants of the different villages considered themselves members of a single tribe, the Beni Isad, and the population of about 11,000 in 1952 was divided among nine clans. At the heart of the tribe, writes Salim, were four clans, each claiming descent from one of four brothers. The brothers themselves were thought to be the descendants of the founder of the tribe of Isad. Accordingly, the four clans held a particularly esteemed position in the tribal hierarchy, as the remaining five clans represented fragmented segments of other tribes who had been incorporated into the Beni Isad by virtue of living among them.

The four core clans were internally ranked, with one, the Ahl ish-Shaikh, being the most respected and powerful. While each clan had its own leader, or sheikh, the leader of the Ahl segment was acknowledged as being the paramount leader of the tribe.

Given this structure, what were the powers of the paramount sheikhs? According to Salim, they ruled over their followers as landowners, military commanders, and judges. Although tribal land was technically the property of the state, or *miri*, the sheikh in practice determined who used it and how much was to be allocated to his followers. Even though the sheikh received a third of the crop, it was not uncommon for him to confiscate more. In the absence of the religious courts associated with urban centers, judicial authority was the exclusive domain of the sheikh. He administered the customary law of the tribe, or *'urf*, with the occasional assistance of an informal council of notables. The executive power of the sheikh was administered by the heads of the

[22] Shaker Salim, *Marsh Dwellers of the Euphrates Delta* (London: Althone Press, 1962).

different clans allied with him, his sons and other close relatives, and finally by a core of retainers and slaves.

The political autonomy of the skeikhs began to erode during the latter part of the nineteenth century. The Ottomans moved army officers and government officials into the area and within a few decades established direct rule over most of the tribes. Indeed, everywhere in the Middle East today, the state's bureaucratic control is important, and instances of full tribal autonomy are hard to find.

How, then, did the tribes and their traditional leaders, the sheikhs, respond to the encroachment of the state? Again Salim provides a good account of tribal politics and its transformation. In pacifying the region the Ottomans attempted to work through the local leaders. While taking strenuous military action against those who resisted, they appointed friendly sheikhs and notables to administrative positions, making them responsible for the safety of travelers, the collection of taxes, and general internal order in the area. Thus, state support became the new basis for the power of the sheikhs, replacing their direct reliance on kin and armed retainers.

In 1924 the political office of the paramount sheikh was officially abolished by the newly constituted government of Iraq; however, the title of sheikh persists and is widely used today as a title of respect. To replace the sheikh, the government established the political office of *sirkal*. Although the *sirkal* serves at the pleasure of the provincial governor, in actuality *sirkals* tend to be members of the dominant lineage of the clans they represent, and sons often succeed to the office of their fathers. Thus the abolishment of the sheikh's political autonomy, suppression of customary law, and intertribal warfare have all contributed to the undermining of tribal independence and the increased integration of these rural populations into the national political system. It has not, however, resulted in the disappearance of tribalism as a charter for local political and social life. It is hard to estimate the saliency of tribal organization in village social and political life throughout the Middle East today. Certainly the exercise of local power and authority using the tribal idiom is inversely correlated with the local strength of the national government in question. In countries such as northern Yemen, parts of Oman, Iran, Iraq, Syria, and Turkey, we find regions where tribal leaders wield great local influence despite state claims and administrative offices. In still other areas where the state administrative apparatus is well entrenched, identification with a tribal name may carry significance primarily as a source of individual and group identity, but it has no direct import for leadership and political power. We take up this subject in a later chapter.

NONTRIBAL VILLAGES

A Turkish village, Sakaltutan, studied by Paul Stirling in the mid-1950s, illustrates the alternative or nontribal form of rural political organization.[23] Sakaltutan, a village in the central Anatolian plateau, was comprised of 100 households and some 600 people. "Here," Stirling writes, "people belong to their village in a way they belong to no other social group."[24] Although this village society is differentiated by distinctions of wealth and prestige, and along lines of family and kinship, there are no sharp class boundaries or clear segmentation along lines of descent. Unlike the tribal peasant of el-Chebayish, whose primary identity is with the clan, the peasant of Sakaltutan identifies with the village or any one of its family-focused political factions.

Sakaltutan, like most Turkish villages, has a territory recognized by the state as its village commons or meadowlands, which is exclusively utilized by the residents. This territory, notes Stirling, is more than simply a matter of administrative convenience; it symbolizes village communal identity. People are willing to mobilize and even fight to maintain the integrity of village communal property.

The formal political structure of Sakaltutan was that established by the state. The village was represented by the headman, who was elected by secret ballot. In addition to the headman, the village was governed by a council of elected elders, who were generally totally ignored by the villagers. Those elected gained neither prestige nor special standing. Rather, when concerted action was required, senior heads of households or large family groups gathered informally to make decisions. According to Stirling, the headman and the council members of Sakaltutan and neighboring communities were young or middle aged. Senior men and individuals of great influence did not hold office themselves but controlled matters from behind the scenes. Shifting coalitions of households within the village formed factions for political action. Although these factions often had a strong basis in kinship, this was not their only organizing principle. More important than kinship was membership in a residential quarter, which included more than one kin group. What politically distinguishes Sakaltutan and other similiarly organized villages in the Middle East is the absence of high-level, overarching, and active descent groups.

El-Chebayish and Sakaltutan represent two contrasting forms of village political organization, but most villages fall somewhere in between, combining features of both. It is a rare village, for example, in

[23] Stirling, *Turkish Village.*

[24] Ibid., p. 29.

which larger descent groups are completely absent or irrelevant. But even where these are found and are politically important, they are frequently cross-cut by factions expressed in terms of residential groupings, party affiliation, and the power of the state.

HOUSEHOLD ORGANIZATION

Tribal and nontribal villages have in common the household as their basic economic and social unit. Termed *beit* in Arabic, *ev* in Turkish, and *hane* in Persian, this group is recognized everywhere as the individual's primary source of food, shelter, and security. Households are economic and residential units usually comprised of people related to each other in a variety of ways. Although unrelated individuals such as servants, hired shepherds, or other retainers may be temporarily part of the household, most commonly those individuals identify themselves with their own families. Whereas patrilineality forms the basis for enduring, named kin groupings, the household may be thought of as a temporary kinship group that utilizes ties of marriage as well as those of descent to define its membership. Women, for example, usually do not lose their identification with their father's lineage or family; however, upon marriage, they join their husband's household or, more commonly, that of his father. This form of residence is termed patrilocal and is nearly universal in the rural Middle East.

Some of the variation exhibited in household organization in any village is due to the uneven distribution of available resources. Households owning much land tend to be larger than those with no or very little land; they frequently include a number of married sons in addition to the head of the household, his wife or wives, and unmarried sons and daughters. Such a household may be termed *extended*, and is the most frequently stipulated "ideal" or preferred form of residence in Iranian, Arab, and Turkish villages. Brothers may elect to remain together in a joint household following the death of their father if their collective property holdings are substantial, and if there are compelling economic interests in maintaining fields and herds intact. Variation in household size and organization is also due to the usual processes by which households are formed, expand, and ultimately break up. These processes can be termed the *domestic cycle*. Sons, for instance, begin life as members of their father's household; after marriage one or more of them may continue to reside in their natal household, making it an extended one. Should their father die during this period, the brothers may elect to stay together, with the oldest assuming the role of head of household. This form of extended or joint household is relatively rare, in

Village men at a wedding feast in southeastern Turkey.

part because it denies at least one brother the status associated with being the head of a household. Most men aspire to head their own households and to assume the social status this responsibility confers.

The household has some social continuity because sons commonly continue to reside in or near their father's household following their marriage. A new household comes into being when a man establishes his own residence after marriage. As children grow, marriages are arranged for sons whereby each brings a bride into the home for some period of time. Married sons most often live in separate rooms added on to an existing house; their common membership in the household is symbolized by their sharing one hearth or kitchen facility.

Although such a household might theoretically continue to grow to include more than two generations of married male agnates, this rarely happens. The large multigenerational *zadruga* households of Balkan villages are seldom found in the Middle East; as noted, sons usually separate during their father's lifetime, and nowhere do we see a consistent pattern of brothers staying together after their father's death. As sons have children, or as frictions arise among members of an increasingly crowded household, men try to found their own domestic units. Thus within any village community, nuclear households are far more numerous than extended or joint households.

The growing reliance of many rural households on income derived from wage labor, with the seasonal or even full-time employment of members in urban centers, further strengthens the trend toward small, independent households. Indeed, a profound social transformation is underway in most of the rural Middle East as the capitalist market economy increasingly penetrates the countryside.

Even though small households are increasingly the norm, many villages usually have one or more households that occupy the most impressive dwellings and that stand out as major contenders for local leadership, power, and influence. They usually comprise a number of closely related adult males, usually father and sons and their dependents. This raises the issue of what the advantages are in maintaining an extended household. One advantage is that such a group provides an efficient means of organizing a reliable source of labor. This presupposes access to such resources as land where coordinated labor pays off. The large household as an economic unit can benefit from economies of scale. The division of labor is potentially quite efficient, even in the context of mechanized agriculture. For instance, one adult son might supervise animal production, while another manages the family's mechanized farm equipment, perhaps driving a family-owned tractor or combine for hire in order to secure additional income. Although mechanization tends to reduce the importance of labor relative to other inputs, skilled agricultural labor is often in short supply. Landowners have to compete with urban employers for the services of skilled drivers and mechanics. It is not feasible to turn over expensive equipment to unskilled operators. Large families possessing both land and capital frequently utilize family members directly in productive tasks. By investing in equipment they can increase the productivity of family members while they reduce their reliance on labor outside the household.

Another advantage of large, extended households is in the domain of politics. With few exceptions, the village dweller in the Middle East must cope with a relative scarcity of good land, water, and pastures. This puts households in a competitive, even a potentially adversary, relationship with each other. Because even closely related households may easily find themselves in competition for land or pasture, the household becomes a fundamental grouping for political action, as it serves to guarantee its members mutual security and acts in concert where property and public image are concerned. Stirling notes for Sakaltutan that neighbors might physically take over land belonging to another, and that there were continual attempts by some to extend their private fields into the village commons.[25] The moving of field markers, diversion of water by stealth, and the trampling of grain fields by animals in search of grazing are some of the problems with which households must contend. Households without active adult males to represent their interests are vulnerable when it comes to disputes within the village. We should note, too, that when machines are involved, the traditional attitudes toward manual labor are changing. Even in well-to-do families, men may operate their own tractors or combines. This reflects what might be called a shift from peasant agriculture to farming.

The point to emphasize, however, is that regardless of the size and composition of households, and despite the many changes we have touched upon, access to resources, social status, and even one's opportunities are largely determined by familial relations.

RURAL TRANSFORMATION: THE CASE OF EGYPT

To better illustrate some of the patterns and processes of change in the rural Middle East, we now turn our attention to Egypt. It should, however, be kept in mind that both for reasons of ecology and politics, Egypt is distinctive. Still, there are relatively good data available and, more important, Egypt was the first country of the region to initiate and experience many of the reforms associated with agrarian change.

Abdel-Fadel, an Egyptian economist, writes that prior to the July 1952 Revolution, owners of large estates or those with over 200 *feddans* (a *feddan* is equal to approximately one acre) made up less than 0.1 percent of the total number of landowners and possessed about 20 percent of the cultivated land.[26] Owners of large and medium-sized estates, or those with five *feddans* or more but less than 200, possessed about 65 percent of the total cultivated land. However, in 1950 44 percent of all rural families were landless. This compares unfavorably with 1929, when only 24 percent of rural households were without land.

Abdel-Fadel points to three major causes for this increasingly uneven distribution of agrarian wealth. First, the big landlords held a monopoly of power over land and water, which allowed them to de-

[25] Ibid., p. 140.

[26] Abdel-Fadel, *Development, Income Distribution and Social Change in Rural Egypt, 1952–1970.*

mand exorbitant rents and dispossess tenants at will. Second, these landowners monopolized the modern credit market, forcing tenants and small landowners to turn to village money lenders who extorted high interest rates, which often exceeded 100 percent per year. These rates often meant the ruin of small landowners and the forced sale of their land. Third, heavy speculation in rural land increased land prices beyond any corresponding increase in productivity, making it difficult for small holders to acquire more fields. (Although the focus here is on Egypt, we must note in passing that similar processes are at work throughout the Middle East.) It is no surprise, then, that land reform was the first priority of the new regime instituted following the Nasser Revolution of 1952.

A survey of two village populations, one from Upper Egypt, the other from the Delta region, provides an interesting village-level perspective that complements the broad economic picture drawn by Abdel-Fadel. Nicholas Hopkins and a research team from the American University of Cairo carried out a study of animal husbandry and household economy using survey interviewing techniques.[27] We rely primarily on their findings from the larger of the two villages, Musha, located in Upper Egypt near the city of Assiut.

Musha is large, even by Egyptian standards, with a population which is officially estimated at 35,000 and which is distributed among some 7000 families or households. Musha can be termed a village, however, because of its limited marketing, crafts, and commercial sectors. The bulk of the households are densely crowded in the nucleated core of the community, although recently there has been some establishment of isolated households in the fields. Members of many households reside either seasonally or permanently in the large cities to the north where they work; others may move regularly among different farming communities as seasonal agricultural laborers. People from Musha have also migrated to Saudia Arabia and Kuwait. The movement of people, both seasonally and as permanent migrants, limited access to land, and patterns of local employment are closely interrelated.

The main crops grown in the village are cotton, wheat, beans, and lentils; animal production is a significant secondary economic activity. The main livestock kept are buffalo, cows, sheep, and goats. Agriculture is heavily mechanized. No animals are used today to raise water to the fields or to till the land. Water is supplied from a main canal throughout the year and is passed through feeder canals until it is pumped up to the fields, where it is conducted by farmer-maintained irrigation ditches to the plots under cultivation. Like many areas in Egypt, Musha has prob-

[27] Hopkins, *Animal Husbandry and Household Economy.*

lems of waterlogging and a rising water table, which have resulted from the extension of irrigation beyond the traditional reliance on the annual flooding of the Nile.

Distribution of land, despite 30 years of land reform and governmental regulation, remains unequal. At the same time, however, alternative sources of employment outside agriculture have opened up and now sustain many households. Although only approximately 10 percent of the village's households are officially considered destitute—that is, without any land or other means of adequate support—a full two-thirds of the villagers are landless. Landless households sustain themselves by finding employment in agriculture, as laborers elsewhere, as shopkeepers, and as civil servants. About 20 percent of the households own land, but of these only 12 percent hold more than 5 *feddans,* or the minimum thought necessary to maintain an average household. Among those with more than 5 *feddans,* a handful have apparently gained great wealth from farming, some of which has been invested in urban real estate as well as in tractors, pumps, and the like.

An apparent paradox is that despite the large numbers of landless, even destitute, families, landlords maintain that there is a labor shortage, that labor is too costly, and that laborers are unwilling to work long hours. In contrast, workers interviewed said that it is hard to find work and still harder to live on the income they might derive from it. Hopkins notes that this debate is still unresolved in the political economy of rural Egypt (and, we would add, in many other parts of the Middle East).[28] Part of the problem has to do with the seasonality of labor demand in mechanized farming. Landowners need to hire at peak season and thus, for a short period of time may compete with one another for workers. The ability of laborers to fully exploit this demand in the form of high wages is limited by the fact that they, too, face competition from laborers coming in from other districts. Though these ramifications go somewhat beyond the report upon which we are drawing, they appear reasonable. Hopkins notes, too, that because workers cannot find employment in the off-season winter months, many migrate to the city, which further aggravates the shortage of skilled farm workers. As noted, the wages and remittances of migrants have to be considered as integral to the village economy today. About 19 percent of the households surveyed by the research team had no direct involvement in farming although they continued to be village dwellers.

Households in Musha vary considerably both in the number of people coresident and how these members are related to one another. While

[28] Ibid., p. 17.

much of this variation has to do with the progression of the domestic cycle itself, some variation is attributable to economic factors, including migration and wage labor. Over 70 percent of the households surveyed were either nuclear (husband, wife, and unmarried offspring) or simply one generational. The survey as a whole puts average household size at 7.6, larger than usually reported for this part of Egypt in the national census. The study notes a significant positive correlation among three interrelated variables: the amount of land owned, large animal production, and household size. Better-off households were larger and more apt to be extended in form. Moreover, animal production was economically more feasible for those with land than for others—the latter being unable to secure adequate fodder or processed animal feeds. In many respects, then, animal husbandry contributes to the wealth differential because it constitutes a form of savings and investment. Young animals are purchased when cash is available, raised, and finally sold in times of need.

We shall not attempt to contrast Musha with the Delta village investigated in the University of Cairo project, but there is one finding that merits comment because it highlights an important aspect of the domestic economy and raises questions that might be usefully addressed in studies in other parts of the Middle East. Approximately half of the women interviewed in the Delta community reported that they worked in the fields, and about 18 percent of them owned livestock in their own right. In Musha, in many respects a more prosperous village, no woman was found (or would admit to) working in the fields. Many did care for animals in their own homes, but still left the grazing and watering of them to the menfolk. The reasons for this contrast are not entirely clear, according to the report, and Musha sounds like an extreme case by either Egyptian or Middle East standards. One reason that might account for the absence of women in agricultural labor is that Musha has a very high rate of outmigration, and many families derive income from remittances. This may enable even relatively poor families to conform to societal values, which stigmatize households whose women work outside the confines of home. Even in the Delta, where this ethic is expressed less strongly, "prosperous households limit the role of women outside the household more than poor ones do—unless that role results from success in education and a government job or its equivalent."[29] These findings caution against assuming that new job opportunities and urban experience for males will radically alter customary patterns of female seclusion and the division of labor in the short run.

[29] Ibid., p. 85.

RURAL TRANSFORMATION: A CASE FROM IRAN

The pattern of internal movement of population and intensification of land use is fairly widespread in the Middle East. In a study of settlement history and land use in the now heavily irrigated Marvdasht plain in Fars, north of Shiraz in Iran, Gerhard Kortum describes a long-term cyclical pattern of village expansion and subsequent decline.[30] Today the organization of villages and towns closely replicates the expanded settlement systems of earlier years. While Kortum notes that settlement expansion is associated with periods of strong central government, he supplies detailed information on the locally limiting factors. Certain environmental constraints and sources of risk for farmers have remained relatively constant over the centuries: Irregularity in water supplies needed for irrigation, soil salinity, breakdown of critical elements in the system of water distribution, competition from nomadic pastoralists, and political unrest all figure in the contraction of settlements at particular times. The most recent low point was reached toward the end of the nineteenth century.

Colonization, nomadic settlement, and investment in water works proceed in times of political security associated with strong state-level administration. Kortum's survey of 356 villages and three urban centers is remarkable for its use of archeological and historical evidence in attempting to see the processes in settlement development that underlie the cyclical expansion of village life in the plains. His thesis is that historically the regulation and distribution of the waters of the Kur River through dams and feeder canals extended settlement into otherwise marginal areas. Inability to control the effects of salinization plus failure to effect appropriate repairs on the critical river dams repeatedly led to the abandonment of previously intensively farmed zones, and the reversion of these lands to pasturage.

Today, the Marvdasht Plain is one of the most agriculturally developed regions of Iran, and in many areas river-supplied water is supplemented by mechanized deep wells which have significantly extended the settlement frontier and transformed village land use. Cotton is now a major cash crop. Industrialized sugar-beet production is another innovation that increases rural cash income.

The effects of capital investment in irrigation are not limited to extending the frontiers of settled agriculture. It has substantially altered the internal organization of production within the rural community and

[30]Gerhard Kortum, *Die Marvdasht Ebene in Fars: Grundlagen und Entwicklung einer Alter iranischen Bewasserungs landschaft* (Kiel: Selbstverlag des Geographischen Instituts

has changed the relationship of local communities to each other and to the marketplaces that develop to serve them. Kortum analyzes the shift in land use and productivity following the completion of the Dariush Kabir Dam in 1972, using data from ten villages whose lands are watered by the dam. Although sugar-beet production was introduced as early as 1935, following construction of the new dam the acreage under beet cultivation increased threefold, and acreage in cotton increased twelvefold. The proportion of land cultivated for food crops or animal production decreased by 42 percent, although the total amount of land put to food crops increased.

In short, diversified subsistence farming has been replaced by an almost exclusively market-directed system of heavily mechanized agriculture. According to Kortum's estimate of agricultural labor, in the preirrigation system of land use, human labor constituted one of the principal inputs or costs. In the new irrigation scheme, labor inputs remain comparable to previous levels at 400,000 people/hours per 1000 hectares under cultivation, while other costs (exclusive of water itself) rise fivefold. This dramatic increase in overhead is accounted for by items that have to be purchased outside the farming communities themselves, particularly fertilizer and machine traction, not to mention the capital cost of canal construction. Cost of labor then shrinks by comparison to become a relatively insignificant component of the agricultural economy.

Although this might be viewed as a measure of the efficiency of this newly commercialized system of land use, it is also indicative of increasing regional and national integration and the dependence of local farmers on distant markets and sources of capital. Commitment to the marketplace and concentration on a limited range of intensively grown crops, such as cotton or sugar beets, puts the small producer, former sharecroppers, and tenants at a serious disadvantage. In Marvdasht, as elsewhere in the agriculturally developed Middle East, such shifts in land use have forced many to seek nonagricultural sources of employment, even when this means migration to the already swollen metropolitan centers.

GENERAL ASSESSMENT

The transformation of agriculture through changes in land tenure, mechanization, and other capital-intensive techniques is by no means uniform or complete even within a single country. In Egypt mechanization is primarily directed at water management, not cultivation; even

A Modern Iranian farmer from northern Iran.

so, the traditional water wheels are used alongside portable gasoline-driven pumps.[31] In other countries mechanization in cultivation and harvesting is more widely employed but not by all, nor is every community experiencing similar rates of investment. Indeed, families and communities are increasingly differentiated in terms of their participation in these processes of change. The introduction of new crops and techniques of production, plus access to new markets are imposing new constraints and lines of social and economic demarcation, just as they offer new opportunities.

As we have noted, the benefits and costs of change are not borne equally. Some of the benefits may accrue to only a limited number of people in a community, while others may be widely shared—for example, better standards of health and education. For many, agricultural intensification has brought the consumer goods and styles of the city within reach. In effect, rural development has somewhat diminished the long-existing gulf separating rural and urban dwellers. Television antennas sprout from village homes, and it is the exceptional household that lacks a portable radio. Cars, trucks, and tractors have largely replaced animal transport and power in many, if not the majority of, commu-

[31] For an insightful evaluation of the role of mechanization in Egyptian rural society, see Alan Richard, "Agricultural Nuclearization in Egypt: Hopes and Fears," *International Journal of Middle East Studies* (November 1981), 13, no. 4, 409–25.

nities. The economic basis for what might be called a rural consumer economy is commercial or market-directed agriculture.

Road transport now integrates most villages with major redistributive centers, thereby giving them access to metropolitan and international markets. Government and private sources of credit underwrite the acquisition of equipment and the purchase of new seed strains and chemical fertilizers. These inputs, however costly, have vastly increased productivity by allowing for shortened fallow periods, double cropping, and higher yields. In short, there is a new rural prosperity evident in the countryside, even in areas like the Nile Delta, which have long been characterized by grinding poverty.

Some of this prosperity, however, is misleading because the costs that sustain it are hidden in sprawling urban slums or in the rows of shacks lining the periphery of an otherwise wealthy-looking village. The size of the farm needed to sustain an "average family" increases rather than decreases as a consequence of integration into larger markets and a more complete reliance on cash or credit to purchase the requisites of production. Families that cannot increase their holdings may fall by the wayside or be forced to move or at least seek part-time wage employment.

In Turkey, Iran, and elsewhere, villages may be virtually abandoned by the younger male populations as the men seek a livelihood outside of agriculture and leave their fields to be tilled by more prosperous neighbors.[32] Some who move to the city do find employment and improved standards of living; many do not. This social dislocation has to be considered as a major cost of the transformation of agriculture and is increasingly a political concern for most governments.

Accelerating rates of land alienation and rural outmigration have stimulated, as Abdel-Fadel notes for Egypt, widespread speculation in land.[33] In most countries there are few safe avenues for investment outside of real estate, including arable land, and few countries have been able to fully control rampant inflation in land values. Even where limits are imposed on acreage owned or on the transfer of land titles given by the government, these can be circumvented in practice. In many ways, these trends in the Middle East seem to recapitulate historical processes experienced in Europe and other industrialized areas of the world.

[32] For a study of urban migration in Iran, see Farhad Kazemi, "Urban Migrants and the Revolution," *Iranian Studies*, 13, nos. 1–4 (1980), 257–77.

[33] Abdel-Fadel, *Development, Income Distribution and Social Change in Rural Egypt*. See also Alan Richards, *Egypt's Agricultural Development, 1800–1980: Technical and Social Change* (Boulder: Westview Press, 1982).

CHAPTER SEVEN
CITIES AND URBAN LIFE

In this chapter we discuss the nature of urban society in the Middle East. To do this we have to ask ourselves what urbanism means and what role the city plays in Middle Eastern culture. In an earlier chapter we said that the cities of antiquity in Mesopotamia, along the course of the Nile, and elsewhere ushered in a way of life with which we ourselves are quite familiar. From that point on, urban-dominated political and economic systems have prevailed and spread, so that today virtually all the people of the world live in the shadow of cities and urban institutions.

In North America and Europe, over 70 percent of the population is urban-dwelling. According to national censuses, almost half the inhabitants of the central Middle East live in centers of more than 20,000 people. Saad Ibrahim, an Egyptian sociologist, estimates that by the year 2000, over 70 percent of the Arab world will be urban—a conservative projection that is likely to hold true for the rest of the Middle East as well.[1]

As we saw in Chapter 1, urbanism in the Middle East, where it all began, meant the emergence of craft specialists, merchants, soldiers,

[1] Saad Eddin Ibrahim, "Urbanization in the Arab World: The Need for an Urban Strategy," in *Arab Society in Transition: A Reader*, eds. Saad Eddin Ibrahim and Nicholas Hopkins (Cairo: American University in Cairo Press, 1977), pp. 361–90.

traders, and laborers whose livelihoods are secured not by farming or herding but by the exchange of services or commodities. Accompanying this is the rise of a cadre of administrators and bureaucrats—that is, a centralized political system that coordinates and even occasionally controls many of the economic activities that make up a complex society. Finally, and perhaps inevitably, as an outgrowth of economic differentiation, we see the development of social stratification and the emergence of social classes. The most striking physical manifestation of these changes is in the nature of human settlement; food-producing communities come to be closely integrated and dominated by towns and cities.

THE NATURE OF THE CITY

The city is more than a concentration of people. Cities are defined by the diversity of functions they serve and their vital roles in networks of communication. A glance at a map of any agricultural area of the Middle East illustrates this point: for example, thousands of villages and many millions of people dwell in a region whose lines of trade and communication are focused on the greater metropolitan area of Cairo, a city of almost 10 million inhabitants.

But size alone is not what is important. In Egypt today many villages number over 1000 households, or 6000 people. Some, in fact, exceed 20,000 in population; one such village has been described by Berque.[2] This village, Sirs al-Layyan, had a population of 22,000 in 1957, but its primary units of political and economic organization were kinship solidarities—large families living in particular parts of the village.[3] This community is a village despite its size—not because of some arbitrary bureaucratic designation, but because of the relatively narrow range of economic and political activities that take place in it. In the less densely populated countries to the east—in Anatolia, Iran, and Afghanistan—settlements of 1000 households would almost inevitably be termed towns because the quality of life in them would be quite distinct from life in smaller communities. There a concentration of 5000 to 6000 people would occur in conjunction with an administrative center, a commercial district, a high school, a police headquarters, hotels, a major mosque, a market, and probably a site of some religious significance, such as a saint's tomb or pilgrimage center.

[2] Jacques Berque, "The Modern Social History of an Egyptian Village," in *Arab Society in Transition*, pp. 183–205.

[3] Ibid., p. 184.

Albert Hourani,[4] writing on the Islamic city, notes that, "A town or city comes into existence when a countryside produces enough food beyond its requirements to enable a group of people to live without growing their own crops or rearing their own livestock, and devote themselves to manufacturing articles for sale or performing other services for the hinterland."[5] Needless to say, those living in the countryside may need considerable inducement to produce such surpluses. Thus some of the services provided by the city traditionally include the collection of taxes, the raising of armies through conscription, and the administering of city-imposed codes of law. Cities and towns, then, are distinguished not only by the marketing and manufacturing functions they serve, but also by their position in a political and economic hierarchy. The city and its hinterland form two interdependent but inherently unequal components.

Robert Adams describes Mesopotamian society, both in antiquity and in the present era, as fundamentally unstable.[6] Part of this is due to the precarious nature of intensive agriculture on this arid plain, and the resulting vulnerability of urban-based regimes and city dwellers to disruption in the supply of vital goods. But at the same time, families or parties from the elite, who are urban dwelling and often culturally dissimilar from the rural inhabitants, come to exercise great power over the countryside, concentrating wealth and erecting monumental buildings to commemorate their power. For the villagers, life has something of a frontier quality. They are caught between the caprices of the landlord or tax collector and the vagaries of weather. However, like frontier people, they have options to exercise when conditions become intolerable, such as taking up animal husbandry and moving away, or moving into the city either to take refuge or to seize the opportunity to profit from periods of urban prosperity. Thus cities and urban life are ever-changing, as political circumstances and conditions in the countryside change.

All this tends to impart a somewhat rural hue to Middle Eastern cities, despite the cultural gulf that may separate the educated or governing classes from the villagers. Cities continually attract rural migrants, who usually live in houses and social settings not unlike the villages in which they grew up. The great Middle Eastern cities today, such as Tehran, Baghdad, Cairo, Istanbul, and Beirut, to name a few, incorporate residential sections that are rural in everything but land use. In

[4] Albert Hourani, "Introduction: The Islamic City in Light of Recent Research," in *The Islamic City: Papers on Islamic History*, 1, eds. A. Hourani and S. M. Stern (Oxford: Bruno Cassirer; and Philadelphia: University of Pennsylvania Press, 1970), pp. 9–25.

[5] Ibid., p. 9.

[6] Adams, *Heartland of Cities*.

fact, most cities were traditionally supported in terms of food by the farm lands around them, and a sizable number of people left the city each day to farm fields beyond the city walls. Urban sprawl and the development of industrial zones changed this for most cities, but in Cairo many urban dwellers—even office workers and small merchants—still retain plots or shares in fields near the city from which they derive part of the food they consume.

Janet Abu-Lughod, an American urban sociologist who has studied Cairo extensively, has called attention to the complex variation found in the Middle Eastern city today.[7] This variation exists not only across an urban-rural dimension where, within Cairo, for example, you may walk in minutes from a fully modern elegant luxury hotel to a rural style mud brick hut without electricity and water, but also across a time/technology dimension: "Side by side stand the modern factory and the primitive workshop, the bank and the turbaned moneylender, suggesting the persistence of a vital residue from yet another variety of urban living."[8]

THE ISLAMIC CITY: ITS STRUCTURE AND ORGANIZATION

Far from a simple contrast between a traditional peasant population confined to the countryside and a modern urban one, careful study of Middle Eastern society reveals a multilayered, complex, and ever-changing social structure. Cairo, for example, combines within it at least three different urban cultures: rural, the traditional urban of the preindustrial city, and modern cosmopolitan. Cairo, like other large cities, may be viewed as a system of subcities. We shall examine the historical transformation and current status of Cairo later on in this chapter, but before we do that, we shall turn to a more general discussion of city and urban life in the Middle East.

When we think of Middle Eastern cities, it is the grand domes and minarets of the imperial capitals—Istanbul, Damascus, Cairo, and Isfahan—that come to mind. Indeed, one often reads of the "Islamic City," sometimes with the connotation of a way of urban life both different from the European experience and possessing an overall uniformity dictated by religious tradition. We cannot ignore that which is unique about the Middle Eastern city. Nor can we deny that urbanism

[7]Janet Abu-Lughod, *Cairo: 1001 Years of the City Victorious*, Princeton Studies on the Near East (Princeton: Princeton University Press, 1971).
[8]Ibid., p. 160.

A view of Istanbul from the old imperial palace. (Photo courtesy of the United Nations/ R. Biber)

has been greatly affected by Islamic civilization. But scholars agree that it is difficult to generalize about a way of life as varied and changing as is urban life in the Middle East. The great metropolitan centers of Baghdad, Cairo, and Istanbul have long symbolized the civilization and cultural accomplishments of the Islamic world. These cities, however, are quite distinct from one another, each reflecting its own unique historical experience. Moreover, these ancient capitals bear little resemblance in their general organization and layout to such new cities as Ankara, Tehran, or Kuwait City.

When Western scholars first sought to understand the nature of these great Middle Eastern cities, they began by comparing them with the city-states of classical Greece or Christian Europe. As a consequence, much has been made of features of urban life found in the West but "lacking" in the East. Among these are the absence of municipal corporations, "free cities," city councils, and "civic mindedness."

The imputed absence of civic spirit is largely a product of European ethnocentricism—as witnessed by the public buildings erected in every Middle Eastern city by private families to grace their communities. Moreover, "civic spirit" is expressed differently in different societies. Besides, this argument ignores the great variability among Middle Eastern cities. Nonetheless, there are fundamental organizational differences distinguishing the Islamic city from those of the West. We can perhaps best appreciate these by looking at how the city in the Moslem world developed and why its institutions took the shape they did.

Foremost among the factors giving a certain degree of similarity to cities in the Islamic world is their shared Islamic ideology and practice, together with the administrative system that took form shortly after the original Islamic conquest. As Stern points out, Islamic law treats all individuals as equal, and as a consequence it does not give special recognition to any group of believers nor to any corporate entities—with only two notable exceptions.[9] The two principal exceptions are the *waqf* charitable endowments and the coresidential family itself.

> *The right of the family to live enclosed in its house led . . . to a clear separation between public and private life; private life turned inwards, towards the courtyard and not towards the street; in the thoroughfares, the bazaars, and the mosques, a certain public life went on, policed and regulated by the ruler, active and at times rebellious, but a life where the basic units, the families, touched externally without mingling to form a civitas.*[10]

The significance of the *waqf* to city life is that this institution provided for the maintenance of most public urban buildings—mosques, great markets, caravanserais, baths, schools, and hospitals—not to mention many lesser edifices fundamental to city life, such as fountains and public water taps. In essence, the *waqf* was a major means by which rural production directly served to sustain urban institutions. The yields from *waqf* agricultural property not only maintained buildings but also paid the salaries of a class of individuals associated with them, in particular the religious functionaries. When a building was erected, its patron would endow it with income-producing property which would

[9] S. M. Stern, "The Constitution of the Islamic City," in *The Islamic City*, pp. 25–50.
[10] Hourani, "Introduction: The Islamic City," p. 24.

ensure its maintenance in perpetuity. This trust is recognized by Islamic law, and such properties have been taken over by the governments or rulers only under near-revolutionary circumstances. In Iran a major source of contention between the Shah and the Shi'a clergy had to do with the monarchy's attempts to seize rural *waqf* land and to control the considerable revenue generated from *waqf*-designated real estate.

By the twentieth century, a substantial amount of arable land in Turkey, Iran, Syria, and Egypt was held in *waqf* trusts. Even though the liquidation of much of this property has benefited the peasants, who were able to get title to fields they formerly worked as sharecroppers, many medieval Islamic buildings have suffered from neglect as their revenues now come directly from the government. As noted earlier, *waqf*-held rural properties were a major means by which rural produce and income were channeled into cities. Further, the administration of these properties by urban-based religious functionaries concentrated considerable power in their hands and made of some of them a privileged class of urban notables.

As for the family, Islamic law recognizes the right of each household to privacy within its own walls. This valued privacy is expressed everywhere in domestic architecture and the use of urban space.[11] Those who could afford to traditionally built houses of several stories around a central courtyard or compound. The lack of external windows helped solve the climatic problems of great heat in the summers and cold in the winters.[12] Streets were narrow and twisted. They were designed to serve the needs of the householders rather than facilitate general traffic. The house itself was traditionally divided between public rooms and private family space—the harem. Poor families could usually not afford this arrangement and often lived in crowded rooms, perhaps adjacent to the home of a wealthy patron or employer.

Prior to the twentieth century, cities in the Middle East were directly administered by the sultan or his local representatives. The city was not a chartered, legal entity, nor were urban citizens distinguished juridically from rural dwellers. However, this does not minimize the importance of cities as administrative or trading centers. Islam, the religion, arose in an urban setting, and its institutions and ritual practices are directed to population concentrations and amenities associated with urban rather than rural life. Its religious leaders, or *'ulama*, simultaneously constitute an urban patrician class as well as one trained in theology, law, and administration. The Arabic word *madaniyya*

[11] V. F. Costello, *Kashan; A City and Region of Iran* (Durham: The University of Durham Center for Islamic and Middle Eastern Studies Publications, 1976), pp. 8–21.
[12] Ibid.

means civilization, refinement, or sophistication, and it comes from the same root as the word for city, which is *madina*.

The first 100 years of Islam saw the conquest of all the lands that are presently Arabic speaking, together with Spain and most of present-day Iran. Within this broad territory lay most of the great cities of Byzantium, the eastern Roman Empire, and the Sassanian Empire of Persia. Of the major metropolitan centers, only Constantinople survived in Christian hands until 1453. Cities such as Damascus and Alexandria, with their established bureaucracies, elites, and great churches and palaces, were taken over and absorbed into the new political order. In A.D. 661 Damascus became the capital of the Umayyad Empire, which stretched from the Pyrenees to eastern Persia, and there is evidence that the existing bureaucracy, at least at the lower levels, was incorporated directly into the administrative structure of this great empire. However, the new rulers were not content to inherit the legacy of the Byzantines and Persians. Rather, they embarked on an ambitious program of urban construction which included the creation of new cities to serve the political, economic, and religious needs of their expanding domain. Indeed, it is hard to appreciate how rapidly and thoroughly Islamic Arab civilization made its presence felt throughout the eastern Mediterranean. It did so through urban institutions and the concentration of skilled administrators and scholars living in cities. Modern urban life continues to build on these social and architectural traditions.

In existing pre-Islamic centers such as Damascus, great mosques were erected, as were schools, palaces, and fortresses. The building of great mosques in Damascus, where a monumental complex was established on the site of the church of St. John, the erection of the Shrine of the Dome of the Rock in Jerusalem, and the construction of thousands of lesser edifices were more than simple expressions of religious sentiment. They were a means of making political statements as they advertised the power and the presence of a new order. They also helped to legitimatize the conquest where it mattered—in the centers of population. Even today, travelers coming from Europe by road find the cultural frontiers of Islam marked by the magnificent domes and minarets of the great mosque complex, as in Edirne (Adrianople), long the principal land route to Istanbul from the West.

Of the new cities established during the early years of Islamic rule, some were garrisons that later evolved into market and administrative centers, such as Basra in Iraq and Ardebil and Qazvin in Iran. Some evolved into major metropolitan centers, and even into world capitals. One, Al Fustat, near modern Cairo, was established as a new residential quarter upon the Arab conquest of Egypt in A.D. 641, but it soon developed into a major city and the capital of Egypt. Its successor city,

Cairo, established near Fustat, now has a population of 10 million, which makes it one of the largest cities in the world. In many respects Cairo is the intellectual and cultural capital of the Arab world. Originally, Baghdad was built as a planned city—one of the earliest examples of a comprehensive urban plan, and for many generations after its founding in A.D. 762, it remained the leading city of the civilized world. Modern Baghdad is on an entirely new site, the old one having been abandoned.

Clearly there are great differences in the organization and quality of life in these early Islamic cities according to how they evolved. But because there is recognizable continuity in the urban tradition, it is useful to mention some of the distinctive features of these urban centers inasmuch as they are relevant to understanding life today. As Costello, Bonine, and others note, urban sites are closely related to the distribution of usable water resources and arable land.[13] The uneven distribution of cities in the Middle East reflects the uneven distribution of these key resources. Further, as Paul English describes for Kirman City in central Iran, urban centers largely determine patterns of land use and production in their hinterlands through the control of capital and markets by the urban elite.[14] Virtually all cities have marketplaces of various sorts where rural goods are sold and urban goods and services are secured.

In addition to marketing, some cities, such as Jerusalem, Mecca, Qom, and Karbala, function primarily as religious centers of pilgrimage, and still others, such as Meshhed in Iran, Konya in Turkey, and Tanta in Egypt, have major religious shrines in conjunction with serving as trade and administrative centers.

Major differences distinguish coastal cities from those of the interior. Cities along the Mediterranean and the Aegean coasts tended to be large and cosmopolitan, reflecting their role as import-export centers serving foreign markets. Most of the African and south Indian trade entered the Middle East via Aden, Jedda, and Basra, and the far-ranging *dhows* carried both Persian and Arab traders to East Africa. Part of this trade consisted of slaves. Today small endogamous black populations of their descendants persist in some countries. All are Moslems and tend to be found along the Gulf coast. Red Sea and Gulf ports remained relatively small in size as they served a far less densely settled hinterland. Until the mid-1950s Kuwait, for instance, had a population of around 35,000, with the main industry being the seasonally active pearl fisheries. Only with oil did this Gulf city emerge as one of the largest ur-

[13] Michael Bonine, "The Morphogenesis of Iranian Cities," *Annals of the Association of American Geographers*, 69, no. 2 (June 1979), 208–24.

[14] Paul English, *City and Village in Iran* (Madison: University of Wisconsin Press, 1976).

ban centers of the region; today Kuwait has a population of about 1 million.

Some coastal cities stand out. Istanbul, Alexandria, and Beirut are international centers, and until recently little of the culture and the quality of life of their citizens was shared by other large cities. These cities had large Christian and Jewish populations and prosperous foreign communities that gave them a cosmopolitan hue. Their strongest cultural linkages, not surprisingly, were to western Europe. Even today, after repeated efforts by nationalist governments to minimize obviously foreign influences, and despite the occasional forced movement of Christian populations, these coastal cities remain qualitatively different. The Italians have left Alexandria and only a few Greeks and Armenians remain, but the sea-side cafés, the special pace of life, and the orientation to the sea and to the Mediterranean way of doing things distinguish this city from Cairo.

We have already mentioned the importance of the household, its right to privacy, and how this affects domestic architecture. However, by and large it is not the residential patterns that distinguish traditional urban life from its rural counterpart. People live in houses and family settings in the cities which, with the exception of the grand houses and villas of the capitals, could be replicated in villages as well. What is distinctive about the city is the complex of buildings and institutions associated with the central mosque, the seat of government, the military installations, and the bazaar.[15] Commonly these are in close proximity to one another; close by, too, are the major schools *(madrasas)*, bathhouses with special days reserved for men and women, the courts, and the governor's compound. Until the administrative reforms of the nineteenth century, the governor's compound did not include a courthouse or federal building. Rather, the residences of officials were simultaneously their governmental offices, much as was the case with the palace of the Shah or sultan, although on a far less grand scale.

The market, bazaar, or *suq*, might be covered, thus forming one vast indoor shopping area, or it might be simply a section of the city's central area devoted to the sale of goods or craft manufacturing. The bazaar area contained hostels for traveling merchants and for workers coming from the villages for seasonal urban employment. Also there would be food sellers, gunsmiths, jewelers, money changers, and the like. Markets dealing in luxury items or expensive manufactured goods such as fine cloth, gold and silverwork, copper utensils, and weapons, then as now tended to be located centrally in the city adjacent to the grand mosque and readily secured with gates and guards. Spices, con-

[15] Costello, *Kashan*, p. 10.

Distinctive architecture of the old city of San'a Yemen.

diments, and imported food items would be sold in this area as well, while food staples, animals, and animal products would be sold in the less centrally located markets, or outside the city altogether.

Within the *suq*, or bazaar, crafts tended to be concentrated in particular streets or buildings, with the cacophonous clamor of the metalworkers announcing their workshops to all. The close proximity of competitors, even when the competitors might be family members, served to generate one of the most valued of commercial commodities: information. Transactions could be conducted with a minimum of privacy, and both buyers and sellers could size each other up. The shopper often returned to the merchant with whom he or she had a history of dealings, perhaps one who extended credit in the past. Because similar traders clustered together, merchants quickly gauged the demand for their items and when bargaining with a customer could rapidly determine the buyer's seriousness of intent as well as the prices offered by the competitors. Bargaining over the price of goods is important for both the buyer and seller. The buyer satisfies him- or herself that the price is right, while both establish or reaffirm a personal relationship that may be useful in the future.

Drawing on his own research on the Syrian cities of Damascus and Aleppo and the extant literature on other medieval Islamic cities, the social historian Ira Lapidus has proposed a general model for Moslem urban social organization during the medieval period—that is, from the late tenth to the fifteenth century.[16]

The basic units of spatial and social organization were the city quarters or neighborhoods, *mehallas*. Their inhabitants were bound together by ties of kinship, common tribal or village origins, and ethnic affiliation. Given the preference for close lineage endogamy, these overlapped. The neighborhoods often constituted distinct administrative units, each headed by its own representative, or sheikh, whose duties included collecting taxes, maintaining order, and acting as the general liaison between the inhabitants of the quarter and the governor of the city.

At certain periods and in some cities, like Baghdad, Cairo, and Aleppo, these urban neighborhoods constituted virtual "walled" minicities that maintained a high degree of autonomy and isolation. There were few panurban institutions that united them. Lapidus writes that the guilds or other professional groupings were very weak and that these were created primarily to meet the fiscal and regulatory needs of the state rather than the membership.[17] These *mehallas* even had their own "police" force—gangs of local youth who acted as militia enforcing public order and defending their urban turf against the encroachment of outsiders. Known variously as *zu'ar* or *fahlawa*, these ganglike organizations persist even today in some cities—for example, in Cairo and Tehran.

Besides the kinship groups and the youth gangs, another important social grouping in the medieval city was that of the religious brotherhood. These fraternal organizations extended beyond the boundaries of the quarter, as their membership also came from the rural areas. Thus they provided an important mechanism for society-wide integration, albeit at an informal and often marginal level.

More significant perhaps, at least in terms of size and formal recognition, were the religious communities that formed around the

[16] Ira Lapidus, "The Evolution of Muslin Urban Society," *Comparative Studies in Society and History*, 15, no. 1 (January 1973), 21–50.

[17] Ibid., p. 49.

religious elite, the *'ulama* and their schools of law, or *madhab*. These schools were organized around a core made up of the religious scholars and their students and their clientele group, consisting of the notaries and clerks in the service of the judges, or *qadis*. Beyond this core, the different schools of law extended to include the population at large who were members of one or another *madhab*. In Lapidus's words, the people looked to the *'ulamas* for

> *authoritative guidance on how to live a good Moslem life, for judicial relief, and for comfort and leadership in times of trouble. More concretely, family ties, the close association of the 'ulama with officials, merchants, and artisans, who were recruited from all quarters and classes of the population, bound the people to the schools and created communities beyond parochial quarters—communities which shared a common law, common norms in family, commercial and religious life, a common judicial authority, and common facilities such as mosques, schools, and charities.*[18]

Although the above information may give us some idea of the "typical" social organization of medieval urban Moslem life, it does not tell us what a Moslem city looked like—that is, its physical shape. What were its typical physical features? And to what extent do these reflect and express the basic social institutions of its inhabitants?

The first and perhaps most crucial feature of a major city would be the citadel—usually constructed on a *tell* (hill) or some other natural defense area.[19] Then there was the royal quarter where the ruler, his court, and entourage were housed. The royal quarter varied in size, location, and significance, but everywhere it constituted more than a palace and attached residential quarters. It enclosed the state's administrative offices as well as the guards and army units. In addition to the citadel and royal compound (which could also be one and the same), an Islamic city would have a central complex shaped along a grid pattern. At the heart of the complex would be the great mosque, or *jami'*, and the *madrasas*. Next to the mosque was the *suq*, or the marketplace, which included workshops, warehouses, and hostels for traveling merchants. The *suq* was arranged so that each major commodity had its own area. Thus there would be the gold *suq*, the cloth *suq*, the spice market, cop-

[18] Ibid., p. 50.

[19] This description of the physical features of the Islamic city is based on A. Hourani's article, "Introduction: The Islamic City," pp. 9–25.

persmiths, and so on. Immediately adjacent to the *suq* area were the great houses of the merchants and *'ulama,* the grand bourgeoisie of the city, beyond which extended the residential quarters of the citizens. As mentioned earlier, these quarters tended to be ethnically homogeneous and fairly autonomous. Christian and Jewish minorities had their own quarters. Finally, there were the peripheral outer areas of the city, where recent migrants from rural areas might well have lived for a while. This was all enclosed by a wall, beyond which lay the cemeteries and the countryside.

CITY LIFE IN THE TWENTIETH CENTURY

To what extent is city life today organized around the physical and social structures of the past? This is a difficult question. Certainly the visitor whose plane has just landed through the noxious haze that frequently envelops Cairo's airport will receive a mixed impression. Having secured a taxi, not always easy to do, the visitor would be convinced by the race through traffic along the divided highway into town followed by seemingly interminable waits in traffic jams that Cairo is both a product and a victim of the twentieth century. Cairo, like the other capitals of the Middle East, is choked with cars, trucks, and buses whose clangor and fumes are appalling. Major sections of the old city have been bulldozed to open up wide roads. Depressingly familiar signs of urban decay and slum poverty abound, and virtually any social ill one might associate with urban slum life can be found in Cairo.

Today all principal cities and towns have been dramatically reshaped to meet the needs of vehicular traffic, and great avenues have been laid out, or simply cut through existing neighborhoods. From Istanbul to Mashed, ancient walls and city gates that have withstood both time and armies have succumbed to urban renewal, while saints' tombs and public cisterns have become foci for traffic circles. In Iran, under the Pahlavi regime, every town was bisected by a new highway or avenue, inevitably leading to a monument glorifying the imperial house. But the city in the Middle East today is more than a network of new streets, avenues, and highways; it is also the expression of new values and a new political order.

Cities in Kuwait, Qatar, and others on the Persian Gulf have been created with the vast revenues of oil, and they show little if any continuity with the past. Kuwait's growth is one of the most spectacular in the history of urban development. As we said previously, at the turn of the century the population of Kuwait was 35,000; serious oil export began after World War II, and by 1957 the population had increased to

Traffic in downtown Cairo (Photo courtesy of the United Nations/B. P. Wolfe)

206,475. This doubled by 1965, and doubled again by 1975, reaching about a million, half of whom are immigrant non-Kuwaitis.[20]

Ankara, the capital of the Turkish Republic since 1921, is a city of grey hues built on a relentless grid of avenues and streets lined with buildings whose bleak architecture emphasized the new republican order. In fact, Ankara's government buildings borrow directly from German models, architecturally announcing a revolutionary departure from the Ottoman and Islamic past. In every Middle Eastern capital and urban center of importance, massive apartment blocks accommodate the burgeoning professional and middle classes, in most cases rapidly replacing both the wall-enclosed villas of the old elites and the handsome townhouses built around enclosed courtyards.

Poor families, everywhere the majority, find housing a major burden. As is common in Cairo, families may share a single room, cooking in the stairwell, using a makeshift latrine, and drawing their water from a public tap serving many buildings. The more fortunate of the urban poor may reside in self-constructed shanties located in districts where rural migrants and others have illegally occupied vacant land.

[20] Saba G. Shiber, "Kuwait: A Case Study," in *Madina to Metropolis; Heritage and Change in the Near Eastern City*, ed. L. Carl Brown (Princeton: Darwin Press, 1973), pp. 168–96.

The Turkish term for this form of housing, *gecekondu,* or "built in the night," evokes both the nature of the construction and the severity of the housing problem that has generated shanty towns around every metropolis, including the new wealthy cities of the Gulf.[21] Although such homes may be superior to the village houses left behind by the migrants, they strike the established town dwellers or middle-class observer as haphazard and dilapidated, and contribute to an impression of urban seediness and decay.

All this is to say that at first glance little remains of the traditional Islamic city except for some sections in such places as Cairo, Damascus, and Istanbul. Beginning in the late nineteenth century, Middle Eastern cities witnessed a major transformation. As was the case in the United States following the American Civil War, the rapid growth of urban centers and their general transformation was not simply the result of new technology and new ideas. It resulted from a radical shift in the regional economy and from the changed relationship of cities to their hinterlands. The American urban revolution resulted from the industrialization of that country and its emergent position as an exporter of finished products. The transformation and growth of the Middle Eastern city was a consequence of the decline of indigenous industry and the increasingly dependent relationship of the region on European sources of capital and technology. Most of the cities that dominate the headlines today—Tehran, Istanbul, Cairo, Beirut, to name a few— took their present form in the context of European mercantile expansion.

Ilhan Tekeli shows how, with the eighteenth- and nineteenth-century dominance of Europeans in Middle Eastern trade, the region ceased being a major exporter of finished products.[22] Instead, it rapidly became a vast market for European goods, with exports largely limited to raw materials. As a result, many inland centers originally developed for trade and as sites of local industry declined rapidly, while coastal cities such as Izmir (Smyrna), Beirut, Tripoli, and Alexandria attracted population and extended their influence, in part through a new merchant class. By the midnineteenth century the traditional crafts and industries experienced near-total destruction by foreign competition.

Historians note that almost overnight the markets of Bursa, a formerly important and thriving town inland from the Sea of Marmara, were flooded with European goods following a trade agreement that opened up Ottoman markets to foreigners. Manufactured towels and shoes, long-famed products of Bursa, were almost immediately replaced by foreign goods imported more cheaply than local craftspeople could

[21] Ibid.

[22] Ilhan Tekeli, "Evolution of Spatial Organization in the Ottoman Empire and Turkish Republic," in *Madina to Metropolis,* pp. 244–76.

make them.[23] These are only a few crafts but they seem to be representative of what happened to Middle Eastern manufacturing in general following the Industrial Revolution and European mercantile expansion. Only now, in the latter half of the twentieth century, has local industry and capital formation begun to compete with foreign products in local markets and then only because of nationalist protectionist tariffs.

CAIRO: A CASE STUDY

Cities cannot be understood in terms of generalities; rather we build our picture of urban life on the basis of specific places and the people who live there. One city we can turn to to fill in the picture is Cairo. Cairo, which we have discussed repeatedly, is, of course, a great city in its own right. But of equal importance, Cairo is viewed by the Arabic-speaking world as the intellectual capital of the Middle East and its paramount political center. As the primary city of the most populous state in the Middle East, it displays all the contradictions and problems of urban life today.

Abu-Lughod's book, *Cairo, 1001 Years of the City Victorious,* is the most comprehensive study of any metropolitan area in the region.[24] Abu-Lughod writes that Cairo, like any city, is more than a complex of streets, subdivisions, and landmarks; it is, in the words of Louis Wirth, a true "mosaic of social worlds."[25] These different social worlds reflect specialized functions and diverse ways of life, which come together to constitute the social and economic fabric of the city. More than any Western city, Cairo, Abu-Lughod suggests, is a place of contrasts and contradictions as it reflects the entire spectrum of Egyptian history and society as well as the revolutionary forces within it.[26] The population of Cairo contains large communities of fellahin recently arrived from the countryside, shopkeepers residing in traditional neighborhoods, industrial workers in government-built flats, and the urban elite in their luxurious apartments overlooking the Nile whose social life revolves around exclusive clubs.

Janet Abu-Lughod organizes her study of "the mosaic of social worlds" in terms of what she calls the urban ecology of Cairo. Accordingly, she identifies "large segments of Cairo where residents share common social characteristics and follow particular life styles that mark them off from residents of neighboring communities with whom they

[23] Ibid.
[24] Abu-Lughod, *Cairo.*
[25] Quoted in Abu-Lughod, *Cairo,* p. 182.
[26] Ibid., p. 182.

CAIRO NEIGHBORHOODS BY SOCIAL CLASS (Note that "Rurban" refers to city dwellers with rural styles of living) Adapted from Janet Abu-Lughod, *Cairo: 1001 Years of The City Victorious,* p. 187. Copyright © 1971 by Princeton University Press. Reprinted by permission of Princeton University Press.

seldom interact."[27] Using census data from 1947 and 1969, she delineates thirteen sections that are distinct in terms of general lifestyle as expressed in dress, patterns of marriage and residence, dwelling types, literacy and education rates, and occupation. Rather than describe all thirteen districts (see map), here we briefly consider three we feel best illustrate the modern juxtaposition of the diverse elements of urban life. The three are: medieval Cairo, entirely encircled by more recent subdivisions today; Imbaba, at the western extreme of Cairo, which at the time of the study contained a large rural population; and the district of Shubra, in which much of Cairo's lower middle class, including a substantial percentage of its Coptic population, resides today.

[27] Ibid., p. 183.

The district of medieval Cairo (area 10 in the center of the map) is essentially all that remains today of the city's medieval heritage. Within it are found some of the most impressive monuments of Cairo's Islamic past, and its covered *suqs* and winding alleys still harbor a way of life that demonstrates a resilient cultural tradition. This is also "exotic Cairo" for the foreigner and the tourist.

The southernmost extremity of medieval Cairo is marked by the citadel built by Salah el-Din, the great warrior who recaptured Jerusalem from the Crusaders. Completed around 1193, it is now dominated by the nineteenth-century mosque of Muhammad 'Ali. At the center of the area, one can still see the remains of the ancient walls marking the site of the original city constructed towards the end of the tenth century at the orders of the Fatimid Caliph Mu'izz el-Din and named al-Qahira (the Victorious). Little remains, however, from the days of its foundation except that neighborhoods, *hara*, of medieval Cairo are still called after the great gates of the ancient Fatmid city, in particular Bab al-Futuh and Bab al-Nasr. The skyline is dotted by the tombs and minarets of the many mosques and tombs built by subsequent dynasties that extended al-Qahira beyond its original walls.

Medieval Cairo is traversed along a north-south axis by the *qasabah*, an ancient street that marked the commercial zone at the heart of the old city. Along this street are the major markets, or *suq*, with their shops, warehouses, inns, and workshops. Each *suq* specializes in a commodity or craft. Here among the gold, cloth, and spice markets and their teeming humanity and clamor, one finds the famous Khan al-Khalili, the tourists' bazaar par excellence. Here, along with the tourist trade, one also finds Cairo's highest concentration of beggars and pickpockets.

Wedged between the silent monuments of Cairo's past and the din and hustle of the busy markets live about half a million Egyptians, many trapped in poverty and neglect. For this is truly an urban slum whose inhabitants eke out a living working here and there in the small-scale trades and informal service sectors that predominate in this area of the city.

At the western edge of Cairo, we find a different and curious mélange which makes up the community of Imbabah (6 on the map). Actually this is less a community than one in the making. The area had once contained large agricultural estates and housed a substantial rural population. Following the land-reform laws, the estates were broken up. Land that was not expropriated by the government was sold to urban developers, and in a short time, both government and private entrepreneurs began to construct buildings to accommodate Cairo's overflowing population.

In 1947 the Imbabah district had a population of 60,000; in 1960 it had grown to around 150,000. Industrial plants to the north of the area attracted enough migrants for a ministry to construct a "Workers City"—a sprawling, low-income project designed to house the factory workers. At about the same time an "Engineers City" *(madinat al-Muhandisin)* was also built to accommodate a number of engineers and draftspeople. All this construction has gradually displaced the original rural population that was concentrated in one fringe area in the west. Thus the community of Imbabah at once houses the hut-dwelling fellah, the apartment-dwelling proletarian, and the villa-dwelling professional—all in close physical proximity. But, as Abu-Lughod observes, this juxtaposition is of a highly transitional character.[28] Certainly by today, none of the rural-type settlements are left in the area.

The third community, that of al-Shubra, lies to the north of Cairo and along a major thoroughfare, Shari' Shubra, which joins the center of Cairo with two modern industrial complexes in the north. In 1960 around 541,000 people lived in Shubra, of which at least a third were Christian Copts. Coptic churches dominate the local landscape and impart some distinction to an otherwise nondescript urban sprawl. An air of genteel shabbiness hangs over the district. Once an agricultural area with a number of palaces and luxury villas, Shubra is today a greyish, monotonous, and undistinguished urban area housing Cairo's petit bourgeoisie.

Nawal Nadim, an Egyptian sociologist, gives a lively ethnographic description of one neighborhood (*hara*) in the medieval section of Cairo mentioned earlier in this chapter.[29] The *hara* studied, that of Addarb al-Ahmar (The Red Alley), contains 117 families with an average family size of 5.5 people. Almost all the families of the *hara* are of the nuclear type, and the twenty-seven extended households were formed by the addition of widows, divorcees, or other single relatives. Although the households themselves are small, the web of kinship uniting them is very far-reaching and important to their economic welfare. Housing is extremely scarce, and most families have found their residences through messages passed on by relatives. Families within the *hara*, particularly the young generation, have intermarried, creating ties of affinity among themselves. Nadim writes that 40 percent of the households reside in one-room dwellings, and that 60 percent do not have water taps and must buy water. Twenty-one families have television sets, which they make accessible to nearly everyone else.

[28] Ibid.

[29] Nawal M. Nadim, "Family Relationships in a Hara in Cairo," in *Arab Society in Transition: A Reader*, pp. 107–20.

The occupations of the *hara* residents may be grouped into three basic categories: *el-mewazafeen, ahl el-san'aa,* and *ahl el-kar. El-mewazafeen* are government employees who have very low status and salary. *Ahl el-san'aa* are artisans and craftspeople whose incomes are the most assured among the three groups. *Ahl el-kar* are unskilled workers, occasional peddlers, and domestics.

Familial and neighborhood interaction takes place within the private and public space of the *hara,* which is at once a spatial and social entity. Physically, Addarb al-Ahmar is a dead-end street or alley with limited access from the larger thoroughfare. Socially, it is a community of households who regard their shared space as their collective private domain. "The passage is used for socializing with neighbors, playing games, raising poultry, cleaning household dishes as well as washing the laundry. It is not uncommon on hot summer nights, to find members of the family sleeping in the *hara* passage by their doorsteps.[30] Even in dress, the residents express a high degree of intimacy and informality that is normally considered as appropriate only within the family.

Although Nadim finds considerable variation in income among the households, and though women in particular compete with one another for status as they display their latest purchases, there is much recognition of interdependence. Men who lose their jobs or face unexpected crises borrow from their neighbors, and there is much sharing of food and utensils among the households. Women, who ultimately control the family's pursestrings, often form lending clubs among themselves, called *gamiya.* Each member of the *gamiya* regularly contributes a sum of money to a common fund which by agreement will go to a particular member once an agreed-upon amount is collected. The members remain together in the *gamiya* until each has had a chance to collect the same sum. This allows poor families access to a sum of cash which would otherwise take them a long time to save. The *gamiya* is an important way in which the poor pool their limited resources and assist each other.

A fictional account of life in a *hara* very similar to the one described by Nadim, written by the famous Egyptian author Najib Mahfouz, captures the human element and the quality of lower-class, old-city, urban culture in Cairo.[31] Mahfouz portrays the residents of Midaq Alley as they cope with their daily lives, and paints a compassionate picture of people who, however poor, pursue lives full of humor, generosity, and occasional drama. The story of Midaq Alley concerns a young, beautiful woman, Hamida, who feels trapped in the narrow con-

[30] Ibid., p. 111.

[31] Najib Mahfouz, *Midaq Alley, Cairo* (Beirut: Khayats, 1966).

fines of her *hara* and longs for the glamour and freedom of life in the "big modern Cairo" outside the alley. Seduced by an outsider who comes to the alley to solicit votes for a political candidate, Hamida runs away with him, leaving behind her fiancé, 'Abbas, the neighborhood barber, her mother, and the hated *hara*. Her seducer turns out to be a pimp who soon forces her to work for him, and she joins the many prostitutes of Cairo. Her fiancé, 'Abbas, had meantime left his barber shop to work for the British Army in order to make the extra money needed to please Hamida and allow them to marry in style. (The novel is set in the 1940s when Cairo was the headquarters for the British Army.)

When 'Abbas returns home to Midaq Alley, he learns that Hamida has left. He refuses to believe the gossip and sets out with his friend Hussain Kirsha to find her. Their search ends in a bar, where they find Hamida drinking with a group of English soldiers. In a fit of rage, 'Abbas sets out to kill Hamida to revenge himself and redeem the honor of the *hara*, but he is killed in the melée. Hamida escapes with few bruises. 'Abbas was a favorite "son of the neighborhood" and the news of his death when brought back to Midaq Alley causes much grief.[32]

Mahfouz's *Midaq Alley* may have been meant to be a parable illustrating the tenacity and "timelessness" of traditional life in the small alleys and pockets of old Cairo. It shows the ways in which the residents cope with and transcend the intrusions and upheavals of colonialism in all its forms and variations. *Midaq Alley* is part of a much larger genre of novels and plays by Middle Eastern authors in which they explore the changing patterns of an urban society, which, like the rest of their world, is in the throes of great upheaval.

These sketches of Cairo are meant to provide some feeling for the organization and quality of life in one major city in the area. We now turn our attention to another pervasive aspect of modern urban life, squatter settlements.

SQUATTER SETTLEMENTS: A TURKISH CASE

In 1961 the Department of Anthropology at the University of Istanbul undertook a field study of what was then a burning political issue in Turkey—the problem of the *gecekondu* or urban squatter settlements. Squatter settlements or shanty towns had sprung up in alarming profusion after World War II. Although particular cities were subject to a great swelling of population due to rural migrants in various periods under the Ottomans, until World War II the newcomers were absorbed

[32] Ibid.

within the existing residential districts. Following the war, however, illegal houses were erected on vacant land in Istanbul, creating new and totally unplanned residential quarters overnight.

The Istanbul University study was designed to answer some fundamental questions regarding the *gecekondu*. Where did the people come from, and why did they choose to settle illegally? Do they intend to return? Are they employed? The Istanbul study was the first of its kind in Middle Eastern cities, and because the squatter-settlement phenomenon is widespread, it is worth describing. Since the original research in 1961, numerous other studies have been carried out in Turkey and other countries, particularly Egypt. We draw on these as well.

The original Turkish study focused on Zeytinkurunu, the oldest *gecekondu* in Istanbul; before the study was completed over 10,000 families were interviewed.[33] Although accurate figures do not exist for all of Turkey, the social significance of the research is indicated by a publication of the Turkish Ministry of Housing and Settlement, which notes that 1965, 45 percent of Instanbul's population lived in *gecekondu* areas. Since then, the growth of the city itself has enveloped many of the older *gecekondu* areas such as Zeytinburunu, while new areas continue to spring up on the expanded outskirts of the city.

As we might expect, given the number of households involved, the research revealed great variation among households within Zeytinburunu. Some families built their own houses, while others rented, and still others purchased houses from informal contractors. Those who owned their homes were soon in an enviable position in comparison to those who rented, as the value of even the smallest dwelling rose precipitously as housing continued to be very scarce. Some areas of the shanty town were well supplied with water and electricity; others lacked even such minimal urban amenities as streets and public water. Regularly, houses were threatened with destruction as municipal authorities attempted to have them bulldozed away.

In spite of variations in wealth and sources of income among the homesteaders, all shared one characteristic: their homes were built on land that the squatters did not own. Commonly, the land belonged to a *waqf* or to the government treasury; some sites were the disputed properties of absentee Christian owners, while others fell within the commons of villages that used to surround the ancient city. The *gecekondu*, then, is a community that evolved in a legal vacuum and in response to problems radically different from those previously experienced by local urban

[33] Nephan Saran, "Squatter Settlements *(Gecekondu)*: Problems in Istanbul," *Turkey: Geographic and Social Perspectives*, eds. Peter Benedict, Erol Tümertekin, and Fatma Mansur (Leiden: E. J. Brill, 1974), pp. 327–61.

populations. This is true for squatter settlements in other countries as well.

The established Istanbul urbanite found *gecekondu* conditions appalling. The village had physically and socially come to the city. Streets were little more than lanes twisting among the houses, turning to knee-deep mud with the fall rains and remaining so through spring. The houses themselves created a ramshackle village appearance, for they were constructed at first of whatever materials came to hand, and only later, as the homesteader accumulated some capital, were they converted into small structures of cement blocks with tile or tin roofs.

The "rural" nature of the settlement, at least in its early days, simply reflected the fact that the vast majority of the families had come from the villages. Moreover, most of the migrants had relatives, or at least people with whom they had close relations as former village neighbors or friends, within the *gecekondu*. Ties based on a shared village or even a common regional background come to be expressed as almost a form of fictive kinship, termed *hemserilik* or "from the same region." We say more about this later; note, however, that from all accounts the social universe which developed initially in the *gecekondu* closely paralleled village or rural models.

At the time of the initial study most households in Zeytinburunu had members employed in factories or as menial laborers, and a smaller number self-employed as shopkeepers or peddlers. Unlike village women, most adult females were involved in one form of wage labor or another, and those not employed in factories knitted clothing at home. Certainly at this early juncture of the squatter-town development, the migrants were heavily represented by families who were agressively seeking ways to improve their well-being and felt quite positive about the opportunities to be found in the city. The high rate of factory employment in 1961 to 1963 generated an optimism and a standard of living far better than that suggested by first impressions of the houses themselves.[34] This, however, was not the case in all *gecekondu* areas, nor was the optimism of the 1960s to last. In 1979 to 1980 high rates of industrial unemployment, together with inflation, caused great hardship in the squatter communities and generated a level of intensive political activity and growing urban terrorism which have shattered most vestiges of the peaceful rural community in the city.

At the time of the 1961 study, a rural quality of life was strongly expressed in patterns of household organization, marriage, and even in ritual. Sons tended to bring their brides home to their fathers' residence after marriage, which was most likely arranged by the parents in tradi-

[34] Ibid., pp. 343–44.

tional fashion. Bride price continued to be paid to the father of the bride. The marriage ceremonies of the village were maintained and the bride was taken from her home accompanied by customary music provided by flute and drums. The wedding party, which in the village would be conducted over several days with guests put up in homes of friends and relatives, soon changed in the city so that the celebration focused on a "wedding parlor" where the groom's family feted the guests because homes in the *gecekondu* were too small to accommodate many guests.

Do the residents of Zeytinburunu wish to return to their villages? "No" responded 94 percent when asked in the 1963 survey.[35] Their answers indicated that for most, standards of living were better in the *gecekondu* than in the home villages; most had little land and no prospect of access to a sufficient amount. A comparison of monthly per capita income of Zeytinburunu with that of four other rural areas of Turkey showed that, on the average, in 1965 Zeytinburunu residents had incomes two and a half times higher than those of their rural counterparts. Most, too, had high aspirations for their children and placed a high priority on education for both their sons and their daughters.

What characterizes the overall organization of Zeytinburunu, and *gecekondus* in general? Even though each varies from the other in terms of physical amenities, access to urban employment, and even resources available for home construction, one common feature is that they are spatially organized along regional lines and by sectarian affiliation. In short, ethnicity is expressed in the composition of neighborhoods or quarters. Alan Dubetsky's 1970 to 1971 study of a *gecekondu* he calls Aktepe, located near Zeytinburunu, shows how ethnicity is expressed, and indicates some of the changes taking place in the older shanty towns.[36] Dubetsky's research is worth citing because, like the earlier study, it draws attention to general patterns of organization evident in other cities in the region.

Aktepe has a heterogeneous population, including Sunni families from the Black Sea and Alevi or Shi'ia families from the eastern mountainous region of Erzincan. The distinction between Sunni and Alevi masks considerable variation within each and forms the basis, Dubetsky suggests, for two social worlds, or "moral communities."[37] The fact that patterns of migration to the city were strongly regional, with relatives joining relatives who had preceded them, gives the *gecekondu* the aspect

[35] Ibid., pp. 358.

[36] Alan Dubetsky, "Kinship, Primordial Ties and Factory Organization in Turkey: An Anthropological View," *International Journal of Middle Eastern Studies* 7, no. 3 (July 1976), 433–51.

[37] Alan Dubetsky, "Class and Community in Urban Turkey," in *Commoners, Climbers, and Notables*, ed. C. A. D. Van Nieuwenhuijze (Leiden, E. J. Brill, 1977), pp. 360–70.

of a series of closely integrated quarters. Clusters of residences are united by common ties of kinship, region of origin, and religious practice, and they usually share one mosque as a congregation. In no sense do we see the radical breakdown in primary ties based on kinship, or *hemşerilik,* mentioned earlier. Even in finding jobs and in the organization of small factories and workshops, such ties are important.

Turkish industry is organized predominantly in terms of family-run firms with a highly personalized system of authority and labor recruitment.[38] The more prosperous of the *gecekondus* are those, like Aktepe, where small-scale manufacturing concerns have developed, drawing on the availability of labor. In Aktepe, Dubetsky counted some twenty-two small shops producing metal products, sixteen woodworking shops, forty-four auto-repair shops, and eleven workshops producing plastic or tile products. There were also eight small factories and five large woodworking concerns. The processes of labor recruitment and the management of the small factories and shops illustrate some of the fundamental organizing principles of the *gecekondu.* In drawing attention to these, Dubetsky also points out what were to become sources of major social and political problems of the following decade in Turkey.

Dubetsky found that laborers and artisans within a workshop or factory are usually drawn from one area of the country, usually that of the owners or managers. Some are their kinspeople, whereas others are likely to be related to other employees. The owner or manager is considered a "patron"—that is, more than a boss—as he is responsible for the social and spiritual well-being of the employee. All share a heavy reliance on what Dubetsky terms "primordial ties," or highly personalized relationships of kinship, *hemşerilik,* or shared regional or village bonds, and close friendship.

The owner, or patron, not only recruits on the basis of a personal relationship but may pay salaries adjusted for such nonwork factors as whether the worker is married, or has pressing personal problems and extraordinary expenses. Workers and their patron almost always share a common sectarian affiliation, Alevi or Sunni, and the system of reciprocal relationships extends into most areas of urban life. The patron would likely assist in medical emergencies, mediate problems with urban authorities, and help in locating housing. Some workshops and factories practice communal prayers, forming a common congregation uniting supervisors and workers. Unlike what has been described for the American ghetto, Aktepe displays a strong sense of interfamilial or community cohesion, perhaps because so little direct social assistance is provided by the municipality. Close personal bonds and

[38] Dubetsky, "Kinship, Primordial Ties and Factory Organization," p. 434.

networks of friends (*hemşeri*) and relatives compensate for weakly developed employment agencies, unions, and welfare organizations.

The picture that emerges is a useful corrective to the sometimes overwhelming first impression of many Middle Eastern cities, an impression of urban poverty and chaotic ramshackle housing. The shanty communities are often markedly better in terms of the basic facilities and general standard of living than the villages where most of their inhabitants were born. What we must bear in mind is that the quality of life as perceived by the poor urban dweller changes both as general economic conditions change and according to whether he or she is able to take advantage or urban amenities.

Dubetsky found a reliance on personalized relationships in the workplace, but he also found that such professions of common interest and mutual support masked considerable exploitation. This dependence slowed the growth of labor unions, contributed to low wages, and only thinly distinguished the strongly divergent interests of those who ran the factories and those who worked in them.

Such urban problems as street crime and civil disorder, and such social pathologies as alcoholism, drug abuse, and delinquency, which a generation ago were virtually unknown in Middle Eastern cities, even in the poorest *gecekondu*, are today the source of constant discussion in Turkish newspapers. In short, the Turkish city, for better and for worse, closely resembles its Western counterparts.

Since the period of Dubetsky's research, Turkish cities have been continually rocked by widespread civil unrest, and terrorist bands of varying political affiliation have seized control of many of the *gecekondu* areas. Turkey is, of course, not alone in this development. We have only to look to patterns of American urban decay and unrest, or to the terrorist assaults on Italian businessmen to see parallels. With the possible exception of the major oil-producing states, the urban centers of the region will likely follow the unfortunate course of urban history in the West.

PLANNED CHANGE: BAGHDAD

Where a measure of public planning and control is present, these urban problems are at least partially alleviated. For example, the Ba'athist government of Iraq has devoted considerable energy and resources to the development of Baghdad and more recently to Mosul, Sulaimaniya, and Kirkuk. Baghdad used to be ringed by a group of shanty towns, called *sarifas*. These mud and palm-frond thatched "villages" were hastily constructed by rural immigrants from the south, most of whom

Downtown Baghdad.

began their migration following World War II. Plans to do away with the *sarifas* had been proposed under the monarchy, but it was not until after the revolution of 1958 that a new planned community, *Madinat al-Thawra* (Revolution City) was constructed to house the squatters who were moved from their shanty town and settled into the new community. Today Revolution City is a clean, bustling town complete with its own schools and municipal services, among which is an efficient and inexpensive bus line that links it to the center of Baghdad, allowing the men to commute to work.

The majority of the migrants who live in Revolution City are rural in origin—peasants who have moved into the city to escape the grinding poverty of the village, the exactions of the landlord or money lender, and the harsh conditions of agricultural labor in the south of Iraq. A study of migrants from the 'Amara region, conducted by A. Azeez, makes it clear that the move to Baghdad resulted in an improved standard of living for the majority, some of whom had become successful shopkeepers and grocers. All informants mentioned that among the attractions of Baghdad were its education and health services as well as its electricity and water supplies.[39]

The newly arrived migrants to Baghdad usually begin by living with relatives and in time, as they find jobs, they move out to places of

[39] A. Azeez, reported in Costello, *Kashan*, p. 43.

their own, not far from their kinspeople and fellow tribespeople. Throughout this transitional period, the new immigrants are helped by a group of fellow tribespeople who act as an informal organization to assist them in their new environment. Azeez points out that loyalties to tribal sheikhs are maintained in the city and, paradoxically, may even be strengthened. Whereas in the countryside the tribe tended to be dispersed over a wide area, in the city they are brought into closer daily contact and mutual dependency. The close proximity of different groups of migrants and city people reinforces the sense of ethnicity and makes the immigrants rally around their traditional leaders, the sheikhs, who now assume new roles as spokespeople and brokers for their group vis-a-vis the government bureaucracy. Neighborhood mosques and coffeehouses serve as meeting places and public forums for the migrant community.

This seeming paradox of the resurgence of "tribalism" in the city has been reported from other parts of the world, most notably Africa, and is the subject of much study and debate. No longer viewed as simply a case of the persistence of primordial sentiment and cultural atavism, it is best interpreted as a form of adaptation to the demands of the city and the modern state. The reaffirmation and politicization of ethnicity (of which tribalism is a variant) in the urban setting is one strategy of adjustment adopted by groups as they compete for resources in a new arena—namely, the modern city.

SECTARIANISM IN BEIRUT

Whereas the 'Amara migrants in Baghdad grouped themselves along tribal lines, Lebanese rural migrants to Beirut do so along sectarian lines similar to what we have seen in the shanty towns of Istanbul. In the early 1970s, Fuad Khuri studied the changing social organization of two suburbs in Beirut, Shiyaḥ and Ghbayri.[40] Until 1956 both were considered one municipality, which was known as Shiyaḥ. The inhabitants were predominantly Maronite Christians, among whom lived a small community of Shi'a Moslems. The increased migration of Shi'a into the area, which began in the 1930s, put a strain on the services of the municipality and resulted in open competition and conflict among the residents, who began to reorganize along sectarian lines. This eventually led to the formal split of Shiyaḥ into two municipalities, Shiyaḥ and Ghbayri. Shiyaḥ was Maronite, whereas Ghbayri was predominantly Shi'a. Sectarian loyalties are expressed and reinforced in public

[40] Fuad Khuri, *From Village to Suburb: Order and Change in Greater Beirut* (Chicago and London: University of Chicago Press, 1975).

ceremonial and organized activities. Shi'a participate in the yearly Passion play that commemorates the martyrdom of their Imam, Husein; the Maronites join the paramilitary Phalange or Kataib party and support its activities.

Khuri interprets the transformations in the previously integrated Shiyaḥ in terms of a gradient. He argues that sectarian loyalties in the urban suburb represent a stage in the transfer of identity and allegiance from the family-focused ones of the village to the national-level ones of the Lebanese state. It may be true that village-level politics and factions are organized along familial groupings; however, where different confessional groups live side by side in villages, sectarian cleavages are equally important there. Moreover, the Lebanese polity itself is organized along sectarian lines, and as the current and still far-from-settled civil war indicates, both the significance of sectarianism and the future of Lebanon are closely linked. The point to remember is that sectarianism in Lebanon should not be viewed as a transitional stage halfway betwen familism and nationalism, but as a fundamental source of social identity and the most important principle of social organization of both rural and urban groups.

GENERAL OVERVIEW

As noted, one of the most salient facts about the Middle East today is its rate of urbanization. In 1970 the Arab-speaking population of the region numbered around 122,600,000, of whom about 35 percent lived in cities and towns of 20,000 or more.[41] This makes for an urban population of over 42 million. Given the present rate of urban growth, however, it is projected that the urban population will double by 1985 and will reach 170 million by the year 2000. For that same year, the total population of the Arab world is projected at only 250 million, which is to say that the rate of urbanization is far outstripping that of overall population growth, and that by the year 2000, all things being equal, the Arab world would be 70 percent urbanized!

The urban growth of the oil-rich Gulf states is perhaps the most phenomenal, both in its scale and speed. Whereas the Arab world as a whole is urbanizing at a rate of 4 to 4.5 percent annually, the rate of urbanization in the Gulf area is 15 to 18 percent per annum; and in Kuwait and Qatar more than 80 percent of the total population is urban dwelling. What is most fascinating, however, is that the Gulf area is on its way to becoming a series of national city-states in which the indigenous

[41] These figures and much of the discussion that follows is based on the article by Ibrahim, "Urbanization in the Arab World."

population makes up only a fraction of the total population, but owns and controls most of the resources and wealth. The bulk of the population is made up of immigrant ethnic groups organized in a semicaste fashion. Thus in Kuwait, the largest city-state in the Gulf area, less than half of the population are natives and full Kuwaiti citizens; the rest are immigrants arranged in an occupational hierarchy that ranges from the professional-class Palestinians, Lebanese, and Egyptians through the Indians, and down to the unskilled Baluch and Pakistani laborers.

But fascinating as these new Gulf societies may be for the scholar interested in urban processes or ethnic hierarchies, they represent a small fraction of the total population in the Middle East, numbering as they do only about 2 million. Besides, their extraordinary oil economy sets them apart. What is more significant are the processes of urbanization in the rest of the region, especially in those countries where the demographic weight is found: Egypt, Iran, Turkey, and to a lesser extent, Syria and Iraq. What are the factors involved in the urban evolution explosion? What are its implications? In considering these, we must keep in mind that the Middle East shares much with the rest of the Third World countries, that it exhibits peculiarities of its own, and that the two are often difficult to separate.

As has been implied so far in this chapter, the rapid increase in urban growth is due to the twin processes of migration and overall natural increase in population. It is not easy to assess the individual contribution of each of these two factors since, as Costello points out, there is no uniform pattern throughout the area.[42] With the exception of the cities of the Gulf, however, most growth is due to natural increase in urban population. It should be pointed out that in the Middle East, unlike other regions of the world, urban and rural fertility rates tend to be about equal, whereas the mortality rate tends to be lower in the cities. This results in a higher growth rate for the cities.

Moreover, as Costello points out, migration and natural increase are also closely tied together because

> *the sequence of events for many major cities appears to involve firstly rapid growth resulting from migration and then within less than a generation a much higher rate of natural increase in the city than in the rural areas because of the youth of the immigrant population and lower death rates. Migration in absolute terms is important in Baghdad, Tehran, Cairo, and other major cities, but it is also an important primer for future urban natural increase.*[43]

[42] Costello, *Kashan*, p. 33.

[43] Ibid., p. 33.

What are some of the social and economic implications of urban growth for the countries of the region? At the present stage and given the lack of planning and control, these implications are largely negative and pose a series of major problems, which include overurbanization, poor housing, declining services, and social unrest. *Overurbanization* is a term used to describe a situation in which more people live in cities than can be properly employed there. Urban inhabitants in the Arab world, for example, consume more than twice what they contribute to the GNP of most countries.[44] In other words, economic development, especially industrialization, lags behind urban growth. Ibrahim notes that in 1890, when Switzerland was on the verge of its period of rapid urban growth, some 35 percent of the Swiss population resided in cities and 45 percent were employed in industry. In Egypt in 1970, by comparison, some 45 percent of the population lived in cities, but only 18 percent were employed in industry. Urban unemployed or underemployed put a great strain on a government's budget and public services, which results in severe housing shortages and the creation of inner-city slums and suburban shanty towns.

[44] Ibrahim, "Urbanization in the Arab World," p. 357.

CHAPTER EIGHT
SOURCES OF SOCIAL ORGANIZATION: KINSHIP, MARRIAGE, AND THE FAMILY

In this chapter we outline what we feel to be the major elements of social organization as based on kinship and family structure in the Middle East. Given the diversity in ways of life and cultural heritages, our discussion is both broad and subject to frequent caveats and qualifications. We are convinced, however, that to understand Middle Eastern society, even its higher levels of political organization, one has to understand the nature of the primary groupings into which the individual is born and how men and women subsequently fashion and use "primordial" ties and relationships throughout their lifetimes. The relationships are perceived and expressed in the idiom of kinship or closeness (in Arabic *qaraba*). Thus, we discuss in some detail the nature of kinship in the Middle East, including patterns of marriage, residence, descent rules, and household organization.

There have been many attempts to characterize Middle Eastern social organization in terms of paradigms unique to Middle Eastern society, as if this region constituted a homogeneous and relatively unchanging world of its own. To that end, scholars have sought some simple and unique code that would reveal and explain social organization in the area. It is as if behavior in the Middle East cannot be understood in the same terms and concepts applicable to Europe or the rest of Asia. For example, an inordinate amount of scholarship has been devoted to the expressed preference for marriage with the father's brother's daughter (FBD). For many scholars, this rule and its extention is a metaphor for what distinguishes Middle Eastern society in general.[1] Other scholars stress the universality of the patrilineal segmentary system in which fundamental building blocks of society are formed according to rules of descent in the male line. For still others, the code lies in a unique system of values whose core lies in deeply rooted assumptions and attitudes about human sexuality and the need to control female sexual behavior. Some trace the origin of this value complex to the Islamic heritage of the area; others relate it to longer-standing practices that revolve around basic notions of honor and shame. Although simplification is inherently attractive, and indeed is the legitimate goal of scholarship, it has to be accomplished with full cognizance of the variability and complexity of the phenomena observed.

When viewing communal organization throughout the Middle East, and indeed throughout the world, there is no doubt that the single most crucial factor underlying social relationships is that of kinship. Even in the most developed industrial societies, most individuals grow up with, coreside with, and, in general, spend most of their time in association with people to whom they are related in one fashion or another. How people interact with one another, their expectations of behavior, and their responsibilities to others are all heavily influenced by whether they are related to one another, and if so how. This specific aspect of social organization, kinship, and the formation of groups on the basis of selected forms of kinship is our initial concern here.

TERMS OF KINSHIP

Let us begin by asking the most elementary question: How do people themselves define and classify their relatives? We have to keep in mind that such systems of classification vary greatly around the world, and that anthropologists have reduced them to a number of major types according to how people define and distinguish key relationships—for example, parents, aunts, uncles, and cousins. This is more than an exercise in comparative linguistics; how people create a specific universe of relations is closely related to patterns of actual behavior. For example, the

[1] J. Cuisenier, "Endogamie et exogamie dans le mariage arabe," *L'Homme*, 2 (1962), 80–105; R. Murphy and L. Kasdan, "The Structure of Parallel Cousin Marriage," *American Anthropologist* 61 (1959), 17–29; R. Patai, "The Structure of Endogamous Unilineal Descent Groups," *Southwestern Journal of Anthropology*, 16 (1965), 325–50.

fact that in English we do not distinguish verbally between maternal and paternal cousins is indicative of the social fact that all cousins tend to stand in the same formal relationship to the speaker. In Arabic, on the other hand, as in Persian and Turkish, the speaker not only has to distinguish linguistically between the maternal and paternal cousins, but also has to specify the gender of the cousin in question and his or her exact relationship to the parent in question.

This descriptive and highly specific system of kin terminology, often referred to as the *Sudanese* type, indicates the significance of distinguishing among cousins in most Middle Eastern societies. This same system distinguishes sets of aunts and uncles from each other in terms of their links to the parents; for example, different terms are used for the mother's sister and the father's sister. Thus an Arabic, Persian, or Turkish speaker may utilize as many as sixteen different terms or combinations to describe immediate blood relations. This precision in referring to people is associated with the importance that is attached to distinguishing among different sets of relatives.

It should be kept in mind that kinship classification systems only reveal part of the structure of social relations; they do not necessarily reveal anything of the actual content. Thus if a Turkish male refers to one man as his father's brother, *amca*, and another as his mother's brother, *dayi*, this tells us only that these two individuals stand in different social points to the speaker. Although in both urban and rural Turkish society, one's mother's brother is generally thought of as being warm and emotionally sympathic to his nephew or niece and one's father's brother is frequently associated with exhibiting parental authority and discipline, it does not necessarily follow that actual behavior reflects these normative expectations. Behavior in Turkish society, as elsewhere, is tempered by matters of personality, expediency, and the specific context under consideration. However, the normative patterns of behavior associated with specific terms do at least set the formal frame and some of the limits for social interaction. Knowing that the terms for close relatives can be extended to more distant relatives—even to strangers under certain circumstances—gives an insight into the expectations that people have of that specific relationship. For example, in the Arabic-speaking world (and in Turkey, too), the use of the term for father's brother, *'ammi*, by younger men to address older men indicates respect and deference to their authority regardless of the actual biological relationship. (Parenthetically we may add that if used in an inappropriate context, the term *'ammi* may indicate derision or patronizing. For example, a wealthy man might address a menial servant of his as *ya 'ammi*, or "my uncle"; in this case, the term expresses the effort of the rich and powerful master to "dilute" the social gap between him and his servant.) Arab children are also encouraged to use the term *khala*, or mother's sister, when addressing their mothers' female friends of similar status and age. This is to indicate both respect and yet informal intimacy.

Extending the terms *'ammi* to a large circle of older males and *khala* to a large circle of older females, while at the same time restricting the usages of *'amma*, or father's sister, and *khal*, or mother's brother, to a narrower group, reflects fairly pervasive social expectations about patterns of cross-generational deference, authority, and even intimacy.[2]

In Turkish, brothers and sisters refer to each other by terms which indicate relative birth order and, by implication, express relative expectations of authority and deference. Naturally, within any household, actual relations between brothers and sisters may bear little resemblance to a hierarchy based on age. But still, there are times when this normative ranking can or should be expressed. For example, at ceremonial meals, or when strangers are present, or when married siblings pay social visits to each other's houses, it is expected that deference will be paid to older sisters and brothers. In a discussion of family structure in a Turkish town, Paul Magnarella notes that older sisters, termed *abla*, become like second mothers to younger siblings, while older brothers, *agabey*, are in many respects second fathers.[3] Older brothers can become quite tyrannical in their behavior towards younger sisters as they assume the guardianship of family honor. Of course, even in a particular community, people from different social classes, educational backgrounds, and the like behave differently. However, in particular societies, it is possible to identify sets of expectations regarding proper behavior towards kinfolk. The way these expectations are met or not met in behavior can be revealing of the nature of the actual relationship between the members of the family.

Thus a basic set of Arabic terms for relatives are understood by the same 200 million speakers of that language, although this vast population encompasses a great range of social diversity and ways of life. All the same, terms take on different values and meanings in different areas and contexts. The same word can convey different values when used in different situations. For example, the Arabic word *'amm* is used to refer to one's father's brother as well as to address one's father-in-law, regardless of consanguinity. In certain situations, as previously described, it is also employed to signify respect for male elders or to

[2] Paul Magnarella, *Tradition and Change in a Turkish Town* (Cambridge, Mass.: Schenkman/John Wiley, 1974), p. 91.
[3] Ibid., p. 91.

Married couple working together at home in a small town in upper Egypt. (Photo courtesy of Diana de Treville)

minimize social distance. Further, one can tease one's age mates by calling them *'amm*.[4]

Husbands and wives refer to one another in a variety of ways that reflect not only social class, but also who is present, whether or not they are addressing each other in private, whether or not they have had children, and other factors that vary with local usage. The basic terms available consist of personal names normally to be used only in private or in intimate conversation, except among the educated upper-class urbanities, where personal names are more widely used in address now. In public, a common practice among all segments of Arabic-speaking society is for spouses to refer to one another (and be referred to by others) by the name of their eldest son—for example, *abu'Ali* or *umm'Ali,* that is, "father of 'Ali" or "mother of 'Ali." If the couple has only daughters, the spouses are addressed by the name of the eldest girl until a son is born. Anthropologists call this practice of addressing parents in reference to their children *teknonymy*. Its presumed significance in the Middle East is to highlight the importance of male issue to both the parents and to emphasize the responsibility of parenting in the marital bond.

Another system of address that is also widespread is a spouse's referring to the other by the kinship term, *bint 'amm* or *ibn 'amm* (FBD/FBS), particularly when they wish to stress their immediate ties to one another, rather than those arising from parenthood. In rural areas in much of the Arabic-speaking Middle East husbands and wives very often

[4]Emrys Peters, "Aspects of Affinity in a Lebanese Maronite Village," in *Mediterranean Family Structure,* ed. J. G. Peristan, Cambridge Studies in Social Anthropology 13 (New York: Cambridge University Press, 1976), pp. 62–63.

simply refer to one another as *hurmti*, or "my woman" and *rajli*, or "my man"—a usage disdained by educated and status-conscious individuals. Emrys Peters found that the terms used by spouses in Shi'ite and Maronite villages of Lebanon differed in a slight but significant manner that reflected their different views of marriage. Husbands and wives in Shi'ite villages consistently used the teknonymous system of address in private—that is, they referred to each other in terms of their children. The Maronites, on the other hand, used either personal names or kin terms such as *bint 'amm* (FBD), indicating that the spouses viewed themselves more closely identified with each other than with their shared children.[5] It is because actual kinship usage is situational that it reveals much about what people expect from others and how they hope to manipulate or affect the behavior of others.

It is interesting to note here that whereas the Middle East is usually characterized as exemplifying a strong male bias with an emphasis on agnatic descent, this is not reflected in the system of kinship terminology. Whereas in many of the world's patrilineal societies, relatives on the father's side are referred to in ways that indicate a closer relationship than with the equivalent relatives on the mother's side, this is not true in the Middle East. The system of kinship terms used by most people of the area, with the principal exception of the Turkmen, is even-handed and can be considered as indicative of the importance that *all* close relatives may play in an individual's life. In other words, regardless of the political and social alignments of the moment, the core of relatives on which an individual may rely and may even manipulate is bilateral and unbounded. As the Arab proverb expresses it, "Your kin are those who stand with you when battle lines are drawn."

PATRILINEAL DESCENT AND PATRONYMIC GROUPS

Ideas of kinship do more than provide a potential network for individual action. Some forms of kinship terminology sort members of a society into groups of people who interact on a regular basis or for some purpose. Just as kinship can describe a larger, almost open-ended circle of relatives, it can also be used to establish some set of relatives apart from the others. The idea of patrilineal descent is an important principle used to distinguish among relatives and to establish potentially discrete kin groupings or categories.

[5] Ibid., p. 58.

Individuals are considered biologically related in equal degrees to both their father and mother, a fact reflected in the bilateral nature of kinship terms used. Still, special recognition is paid to those relatives with whom one shares a common ancestry in the male line—that is, one's father's lineal relatives. At its simplest, this is directly analogous to the way in which family names are passed on through fathers in American and English society. In general, however, patrilineal descent is much more significant than just simply a means by which family names are passed on. In both tribal and nontribal Middle Eastern societies, a number of important rights, duties, and mutual expectations are associated with close patrilineal kin. In tribal societies, these rights include rights to water, land, mutual defense, and so on, as we have already seen in the earlier chapter on pastoralism. In nontribal communities, an individual may depend on his or her patrikin for protection, economic assistance, and general support.

Fundamental to all Middle Eastern social groupings is the idea that inheritance rights, which often establish an individual's basic access to productive wealth, are agnatically defined. The *'asab* (literally sinew) relationship defines primary heirs as those related to the deceased person in the ascending and descending male line. When political authority is passed through inheritance, it also follows the male line. Throughout their lifetime, women continue to partake more in their father's social status than in that of their husband.

What is important for Middle Eastern society is the use of the principle of patrilineal descent to form relatively small-scale groupings for social action. These can be termed extended families, patronymic groups, or shallow lineages. Although tribal organization is based on the widest extension of this principle, the use of descent is more widespread to define small-scale groupings of the sort that characterize village, town, and city alike. This does not mean that all groups for joint action, coresidence, or whatever are restricted to this patrilineal model. People may form groups in which the closest ties of kinship are through maternal relatives, or they may be joined by a mixture of patrilateral and matrilateral ties. Even when patrilineal groups are recognized, unrelated individuals may still join. For example, this often happens when a family settles in a village in which it has no previous patrilineal ties to the other residents. With time, facilitated by intermarriage, the family and its descendants may well become assimilated into one of the dominant patronymic groups. In short, although the actual makeup of any given social group is apt to be varied, a consistent pattern is to recognize the primacy of the patrilineal ties in forming named groups, be they families or lineages.

THE FAMILY AND THE HOUSEHOLD

Although it is sometimes difficult to distinguish the concept of the family from that of the household, the two are not always coincidental. Families are social constructs based on marriage and consanguinity—in particular, relations of descent. It is this latter relationship that gives families continuity across generations. Households, as we use the term, are the economic and residential units that may or may not correspond with the family. This analytical distinction between family and household is one made by members of Middle Eastern communities themselves, as expressed in the Arabic term *'aila* for a named core of closely related kinspeople and their affines (relatives by marriage), and *beit* or *dar,* which refer to the smaller, coresiding group.

The traditionally ideal household structure among Arabs, Turks, and Persians is comprised of an extended family made up of the father or patriarch, his wife, one or more married sons and their families, and all the unmarried daughters and sons. Although this form is still thought of as the "ideal" by many—rural and urban alike—it probably was never the statistical norm. In 1974, in a sample of 181 households in a Turkish town, Magnarella found that 22 percent appeared to fit the ideal type or a variant of it.[6] In a sample of 171 nomadic Turkish households, Bates found 30 percent to be extended.[7] In the Turkish census of 1965, 27 percent of village households contained two or more related families (married couples), towns contained 14.5 percent, and cities contained 9.8 percent. It is most likely true, as Magnarella suggests, that there is an overall tendency for the number of extended family households to decrease in correlation with the increasing urbanization and industrialization of the country.[8] This is reflected in changing attitudes towards the very idea of extended households; 82 percent of young adults interviewed by Magnarella indicated that newlyweds should live alone without parents.

In a study carried out in 1970 in the city of Isfahan in Iran, John and Margaret Gulick examined two neighborhoods, one a traditional quarter, the other a new squatter settlement on the periphery of the city.[9] The project was designed to document variation in household size and domestic organization, in addition to other factors related to fertility and

[6] Magnarella, *Tradition and Change,* p. 91.

[7] Daniel G. Bates, *Nomads and Farmers: A Study of the Yörük of Southeastern Turkey,* Museum of Anthropology, University of Michigan Anthropological Papers No. 52 (Ann Arbor: University of Michigan, 1973), p. 73

[8] Magnarella, *Tradition and Change,* p. 93.

[9] John Gulick and Margaret Gulick, "Varieties of Domestic Social Organization in the Iranian City of Isfahan," *Annals of the New York Academy of Sciences,* 220 (March 1974), 412–69.

child rearing. The Gulicks found great variability in the makeup of the larger residential groupings, called *compounds;* some consisted of related households, while others consisted of households that were unrelated to each other. The individual households themselves ranged from those containing a simple nuclear family to those with complex extended families with a number of married siblings. In this major city of half a million people, kin relied on kin in arranging their domiciles and in making residential decisions.

A compound consisted of a building whose rooms were arranged around a courtyard and were generally closed to the outside except for a single heavy wooden door. Residents shared a common latrine and occasionally a common brick oven, although cooking was usually carried out separately by the individual households dwelling within the compound. In addition to multiple household compounds, some dwellings were occupied by one family alone. On the average, each compound contained 2.05 households, but some had as many as 5. Household size for the total sample averaged 4.86—a figure we should note because it agrees closely with the average household size as measured in national censuses in many countries in the Middle East.

A household is defined as a family grouping that commonly eats and lives together, generally in the same room. Most of the households in the sample consisted simply of a man, his wife, and their unmarried children. When one considers the compound, however, it is clear that the majority of households are located near others containing close relatives. Thus, many compounds are in essence a form of extended family. A difference between the old neighborhood and the new is that the latter had almost twice the frequency of small, single-household compounds. This, it would seem, reflects poverty and the high cost of building large compounds, not a shift in attitude toward residence. Where families do reside near one another in either neighborhood, they are most likely related by patrilateral ties. This is not, however, inevitable, and a significant number of compounds are formed by households related via the wife, her parents, or her siblings. In conclusion, the Gulicks write that despite the clear tendency for domestic organization in Isfahan to be built upon ties of patrilateral kinship, one must keep in mind the variability that exists. There is a high rate of matrilateral cousin marriage and matrilocal residence, and also people who for various reasons reside alone or with people to whom they are not related.

The household does not determine the limits to actual patterns of social interaction. Very often in rural as well as in urban settings, a group of brothers or agnatic cousins cooperate in particular political or economic enterprises, but reside in separate households—each rearing his own children and maintaining separate household budgets. There is great variability in actual patterns of residence and in household makeup, both in terms of numbers of people and their relationships. But we can safely generalize that the more inclusive social unit of the family, however formed in practice, is important and continues to be so even as residential patterns change. Moreover, the size and significance of named family groupings in the Middle East correlates closely with resource availability. Rich families take pains to maintain close relations among their members and to utilize ties of kinship and marriage to reinforce, perpetuate, or advance their positions. Whereas a poor man would likely find little to draw his own in-laws close to him, a rich or powerful man will probably make use of (and be used by) not only his own in-laws but those created by the marriages of his children. Even where productive property is individually held, it is advantageous for members of wealthy families to act in concert. Thus, when one hears reference to the decline of the family in the Middle East, it is usually in reference to the breakup of large residential entities and not necessarily to the diminution of the viability of the kinship grouping itself.[10]

The great variability in family in terms of size and social significance is not simply a result of urban as opposed to rural society, nor is the contrast one between "modern" versus "traditional," or even tribal versus nontribal. People organize and maintain themselves in large familial groupings and ascribe social significance to them in proportion to the material benefits that accrue from such organization. Where there is individual access to wealth, power, and prestige, those who attain it usually attract and maintain clusters of kin around them.[11] The poor in both rural and urban communities tend to resemble one another in that they are organized around small and unstable family groupings. The powerful in both rural and urban sectors are alike in that they frequently use part of their wealth or influence to develop and maintain large and enduring kin groups.

MARRIAGE

Rich or poor, large or small, all families in the Middle East seek to control marriage, which is viewed as essentially a union between families rather than between two individuals. Marriage is very often employed to

[10] Samih K. Farsoun, "Family Structure and Society in Modern Lebanon," in *Peoples and Cultures of the Middle East*, Vol. 2, ed. Louise Sweet (Garden City, New York: Natural History Press, 1970), pp. 257–307.

[11] Hildred Geertz, "The Meaning of Family Ties," in *Meaning and Order in Moroccan Society*, eds. Clifford Geertz, Hildred Geertz, and Laurence Rosen, Cambridge Studies in Cultural Systems (London: Cambridge University Press, 1979), pp. 315–80.

reaffirm or strengthen existing familial ties as well as to build new ones where none existed before. There is, in fact, a frequently expressed preference for marriages between first cousins and in particular between a man and his fathers's brother's daughter. This latter preference, which is by no means universal, is often described in the literature as a peculiarity of Middle Eastern marriage systems. The early Islamic precedent cannot be discounted in affirming the preference for FBD marriage, and even as Islam spreads today in Africa, this preference for close-cousin marriage often follows. 'Ali, cousin of the Prophet, married Fatima, the Prophet's daughter, quite likely following a well-established pre-Islamic practice. Thus, for some Moslems, such a marriage as a celebration of religious tradition is reason enough to stipulate it as an ideal. The theoretical and structural implications of this marriage rule are taken up in some detail when we further discuss descent and tribalism; at this point we consider the issue only from the perspective of the individuals involved and their immediate families.

To begin with, a word of caution: We must be wary in ascribing undue structural significance to the expressed preference for marrying FBD, as the studies available indicate a wide range of variability in marriage practice, even within communities in which this preference is strongly voiced. Moreover, too few studies utilizing representative samples have been published to allow easy generalization about its prevalence. What we can safely say, however, is that there is a strong tendency for people to marry cousins in general, and that this practice crosscuts most ethnic and class differences and tends to be shared by Moslems and Christians alike.

Rates of cousin marriage vary greatly from group to group and indeed even among different stratas of the same community. For example, in her study of marriage and property in an Arabic-speaking village in southeastern Turkey, Barbara Aswad found that a significantly higher percentage of first marriages contracted by the landowners were with the father's brother's daughter.[12] This was in marked contrast to marriages among the poor and landless. In a sample of 473 marriages among the Yörük of southeastern Turkey, Bates found that nearly 22 percent involved women married to their father's brother's son.[13] About 40 percent of the marriages in the same sample were between first cousins and second cousins of all sorts. These high rates of close endogamy reflect not only the strength of a cultural preference, but also that the Yörük are a relatively well-off ethnic minority in the region where they live.

[12] Barbara C. Aswad, *Property Control and Social Strategies: Settlers on a Middle Eastern Plain*, Museum of Anthropology, University of Michigan Anthropological Papers No. 44 (Ann Arbor: University of Michigan, 1971).

[13] Bates, *Nomads and Farmers*, p. 73.

This practice is mirrored throughout the region among both Christians and Moslems in the tendency for families of high status and groups controlling important resources to marry within themselves. From this wider perspective, marriage with the FBD or another close relative becomes simply an extreme expression of generalized endogamy, the principle that states that people should marry within their lineage, community, village, and social class. In fact, this equality of status is expressed in the *Shari'a*-based marriage rule termed *kafa'a.* Among most Moslems this rule demands that the couple be of the same or equivalent social background, but it does not spell out exactly how this equivalency is to be determined. Even though the rule of *kafa'a* is rarely invoked as a legal impediment to a marriage that has been agreed upon by the families, it does establish the basis for negotiations over the amount of bride wealth or dowry, which is often viewed as symbolic of the relative status of the families involved.

Also important to understanding marriage endogamy are patterns of inheritance, especially among people with real property. As we have seen, the *Shari'a* entitles a woman to a share of her father's property. Among poorer families in most communities, daughters are customarily excluded from claiming their share, but among the wealthy, women do claim what is theirs. Under these circumstances, even though the formation of large patrimonial estates is precluded by partible inheritance, a cluster of close relatives can maintain continuity of control over contiguous plots of land by marrying among themselves. This helps to explain why control over marriage of women is usually a more important issue among the propertied class than among the propertyless. Thus, close endogamy is one more mechanism used by local families of note to perpetuate their position.

Also, in close endogamy, the affinal or in-law ties created go beyond immediate economic expediency. The renewal or reinforcement of existing relationships through intermarriage announces to the community the importance of the particular family as an enduring group. In marrying a close cousin, a man not only expresses his close association with his uncle or other relative, but also affirms the fact that his own father and his father-in-law are on close terms.[14]

Given the importance of affinity in expressing and maintaining social ties, it is not surprising that polygamy, or marriage with more than one wife, is permitted in Islam. The *Shari'a* allows a man to marry up to four wives at any one time, although this provision is couched in a set of restrictions that seemingly limit its practice. Among traditional proper-

[14] Fredrik Barth, "Father's Brother's Daughter Marriage in Kurdistan," *Southwestern Journal of Anthropology,* 10 (1954), 1964–71.

tied classes, particularly in the rural areas, a man might marry a close cousin for his first wife and take a second wife from another family, thus extending his social and political network. Again, as in the case of marriage with the FBD, rates of polygamous marriages vary greatly from one community to another and in terms of social backgound and class. Apart from Saudi Arabia and the Gulf states, it is rarely encountered among urban-dwelling upper- and middle-class families. In fact, modern legislation in many countries (for example, Turkey and Tunisia) has either forbidden it completely or has put impediments to its practice.

Comparative data are still lacking on the practice of polygamy. It seems, however, to be limited in its occurrence to traditional landowning families in communities in which secular education is not very significant—for example, among the great sheikhs of the Euphrates delta of southern Iraq and their wealthy counterparts among the Yomut and Göklan. Turkmen of north Iran almost inevitably keep polygynous households. Increasingly, however, these same families are sending their daughters to school, and polygyny will quite likely decline. Even where it is still practiced, the ability to maintain a polygynous household contributes less and less status to the male.

Monogamy, always the practice of the majority, is increasingly becoming the ideal form of marriage. This reflects social as well as economic changes in the society, among which are the decrease in infant mortality, which encourages families to have fewer children, and the cost of rearing and educating children, which can present problems even for rural people.

The increasing value placed on monogamy also reflects the rapidly changing status of women. No longer simply confined to being bearers of children, women increasingly pursue jobs and attain status in their own right, a development which is incompatible with the personal restrictions inherent in traditional polygynous marriage. In the cities, the increased opportunities for women to be educated and employed outside the home further discourage polygyny, which is increasingly viewed by educated men and women alike as exploitative of women and "old fashioned."

MARRIAGE ARRANGEMENTS

Throughout the region marriage is customarily initiated by negotiations between families, rather than as the outcome of individually pursued courtship. Of course, even in this area rapid changes are occurring, particularly among urban people, where open forms of courtship are increasingly evident. In Cairo male and female students at the universities attend classes together, and there is considerable socializing outside the classroom. As a consequence, marriages among the educated are very much the result of individual choice, though it is rare that any marriage will go against expressed family wishes. This pattern for the educated classes is generally true throughout the Middle East, except perhaps for Saudi Arabia. In Turkey the freedom of courtship is well established even among the less-educated working class.

Despite the increase in individually initiated courtship in some sectors of contemporary society, the prevailing normative pattern remains that of familial negotiation and arrangements. Even when courtship results in a choice of a spouse or even in cases of elopement, the families ultimately go through the process of a formal meeting and agreement. One important part of the negotiations is the nature and amount of the *mahr,* an Arabic term that is widely used by non-Arabic speakers as well, and is usually translated as bride wealth or bride price. The *mahr* remains significant not simply because of its possible economic value, but because it is required by Islamic law to validate any marriage contract. Thus, bride wealth is more than a folk custom carried over by force of tradition.

Mahr involves the transfer of an agreed-upon sum of money or goods from the bridegroom's family to that of the bride. Technically speaking, the preponderance of legal opinion is that the *mahr* should become the property of the bride herself. In practice, where significant amounts are involved, the *mahr,* or a good part of it, often remains with her father or guardian. Again, there is much local variation as well as differences within communities according to social class, education, and even family reputation. In some communities or among some ethnic groups, the *mahr* may be regarded as fixed, and in any particular period, all will pay approximately the same sum. The Turkmen of Iran illustrate this approach with a tradition that bride wealth must amount to the cash equivalent of *ten mal,* or ten horses. Among some rural communities in Turkey, agreement has been reached to fix bride price at a consistent and low amount, and thus to avoid the rampant inflation in the *mahr* which afflicts most of the region. In some areas, as in Iraq, inflation in *mahr* payments has become a national issue debated at all levels of the society.

Even in communities in which a high *mahr* is the norm, some families, usually the wealthy, may make ostentatious efforts to demonstrate that the *mahr* does in fact go to their daughters in the form of jewelry, furniture, property, and so on. Among the urban elite, there is great variation in how the *mahr* is handled. In the large cities of Turkey, there is strong sentiment that the *mahr* be considered as strictly symbolic, with only a token exchange or even none at all. In provincial towns or in villages, however, the *mahr* may indeed involve a very substantial payment. In other countries, for example Iran, the publicly announced amounts of the *mahr* may be greatly inflated because families strive to display their status in this manner. The actual payment, on the other hand, may be only a small portion of the announced amount. Even the precentage of the *mahr* actually given the bride can vary for many reasons. Some families might give less of the *mahr* to a daughter they have put through school or formal vocational training, the implication being that the father is to be compensated for this and for the increased earning power she will bring to her husband.

A common rule regarding *mahr,* if a formal written contract is drawn up, is that about one-third of the negotiated sum is paid upon marriage, and the balance is held as security against the possibility of a future repudiation of the wife by her husband. If this is spelled out in the marriage contract, it can be considered as a form of anticipated alimony. This is common among the elite of the region, except in Turkey. Even when people frankly acknowledge that this sum is unrealistically high or that its collection will be highly improbable, it serves in some circles as an index of social status, or at least pretension. In rural society, this practice is rarely encountered except among wealthy landowners.

Marriage arrangements among rural Moslem populations as described in ethnographic accounts often resemble a straightforward exchange of a woman for money or goods, with little pretext of providing the bride with a comparable trousseau or insurance against possible repudiation.[15] The *mahr* here becomes a bride price, and the family is conpensated for the loss of the woman's labor and child-bearing potential. The money received is used to acquire a bride for a son, or for any purpose a family decides upon.

Bride wealth, or *mahr,* however paid, is never divided among members of the extended family in the Middle East, as it is in many parts of the world to emphasize the collective responsibility of the larger kin groups towards its members. In the Middle East the emphasis is on the primacy of the rights and obligations of the father or his surrogate, and to a lesser extent the paternal uncle toward the woman. In this sense, marriage is primarily the concern of a very restricted familial grouping, rather than the collective responsibility of such larger social entities as the lineage, clan, or village.

One good composite description of the marriage arrangements and ceremony is that provided by Paul Magnarella, again for a small Turkish

[15] See William Irons, *The Yomut Turkmen: A Study of Social Organization among a Central Asian Turkic-Speaking Population,* Museum of Anthropology, University of Michigan Anthropological Papers No. 58 (Ann Arbor: University of Michigan, 1975), for a good account of this among the Yomut of north-central Iran.

town.[16] In general outline the approach is closely replicated throughout our area. As Magnarella describes, much of the ritual associated with marriage emphasizes the roles of a wider circle of relatives in arranging the marriage and the significance of any marriage for groups of people, rather than for the two individuals directly concerned.

Traditionally, and often today, the boy's family initiates the search for a bride among "honorable and reputable" families. Steps here are informal and involve only indirect contacts. The initial steps of the negotiations fall on the women of the two households. Only after agreement has been reached on all the important particulars is a formal meeting arranged, again involving female relatives of both households. At this juncture, if all goes well, the boy's side will send male intermediaries to formally request the girl's hand in marriage and to reach an understanding on the value of the *mahr*.

The next major step is an engagement ceremony, again involving women coming together and exchanging gifts. The bride-to-be symbolically indicates her respect and formal subordination to her future mother-in-law by kissing her hand publicly. As Magnarella suggests, this ceremony stresses the importance of the mother/daughter-in-law relationship, and the presence of the women of the two households symbolizes the union of the two kin groups.[17] Following this, perhaps as long as a year later, is the formal marriage presided over by a religious functionary. In addition to the religious functionary, the participants in this ceremony include the groom, two witnesses, and a representative of the bride's family. In this specific Turkish community, as is often the case elsewhere, the bride's physical presence is not required at the formal marriage ceremony. Both the groom and the representative of the bride are asked three times if they concur to the marriage; upon affirmative answers to each inquiry, the ceremony is complete.

Following the ceremony, the bride is transported with much to-do from her natal home to that of her father-in-law. Even if she is to reside in a separate dwelling with her new husband, she must nevertheless be taken first to her father-in-law's house, which symbolically emphasizes the ties between the two families.

POSTMARITAL RESIDENCE

Residence after marriage follows a regular pattern which can be instrumental in shaping the organization of the local community, particularly where people continue to reside in close association with their

[16] Magnarella, *Tradition and Change.*
[17] Ibid.

Three Yörük brides from one household.

relatives. The primary rule of postmarital residence everywhere in the Middle East is that the bride leaves her natal home to reside with her husband. He, in turn, by custom and practice, usually continues, at least for a time, to reside with his father or in close proximity to him.

Although there is much variation in terms of the actual composition of households and of the neighborhoods they form, the incorporation of the wife into her husband's family or household is virtually universal. It is a rare exception to find a man physically joining the household of his wife's father, although the practice of establishing a separate residence near her kin is frequently seen. It is explicitly felt in almost all sectors of Middle Eastern society that it is somewhat humiliating for a man to reside with his father-in-law because it would indicate that his family lacks social status or sufficient resources. It puts the groom in position of subservience to the authority of his father-in-law. While it is thought of as natural for a son to be under his father's direct or day-to-day authority, this does not extend to the father-in-law, even when the father-in-law is a close relative. Even a man's father's brother who enjoys a generalized position of authority is not normally considered of the same status as one's own father. Furthermore, should the bride's father be forced by circumstances to move into his son-in-law's home, he would suffer a similar loss of status, as this would bring him under the authority of his son-in-law. However, if a widower were to join his son's household, he would continue to occupy at least a nominal position as head of the household and senior male.

As mentioned earlier, there is a widespread feeling that a man should avoid informal contact with his wife's parents in the first years of marriage. Such meetings as occur are likely to be ceremonial meals and visits. This avoidance clearly serves to minimize the potential friction attending the process of the transfer of responsibility for a woman from one male-defined group to another. As we might expect, the period of formality and avoidance is considerably less when close relatives intermarry.

The transfer of authority over women deserves some elaboration, as it points to some aspects of interfamily dynamics. Immediately after marriage, the husband assumes sexual rights, while his father or male surrogate assumes direct responsibility for the general well-being of the bride.[18] In this sense the bride becomes part of another household under the authority of other males of her father's generation. In fact, one often hears, in Arabic or Turkish, the new bride referred to as "our bride" by members of the groom's family. The responsibility for the bride's good conduct and reputation ultimately remains with her father or brothers; only with the passage of time and with the bearing of sons would that responsibility be completely relinquished. In this sense, the movement of the woman into a new family or household is seen as a gradual process of incorporation and not as a single event. As we have said, the woman never loses her identity as a member of her natal agnatic group, and she continues to partake in the social status of her father as much or more than in that of her husband.

CONJUGAL VERSUS DESCENT TIES

Conjugal ties, however inportant for the formation and maintenance of households, do not supersede the jural rights stemming from descent. For example, in the event of the death or divorce of the wife, the ongoing household unit is based on the relationship of a man to his children, or even his grandchildren. Under no circumstances is it expected that children would come under the control of the woman's natal family. Further, when a woman is divorced, upon reaching a certain age (generally 6 years), her children automatically revert to the custody of their father or his male kin. Unmarried children of the same father but of different

[18] An interesting exception is the Yomut Turkmen. A man can expect to cohabit with his wife only some years after marriage. Until then, she continues to live with her father, and the husband has no visiting rights. See Irons, *The Yomut Turkmen.*

mothers usually reside together and maintain close ties, being of one family. Children of different fathers but sharing the same mother rarely reside together and may not, in practice, recognize a close relationship unless their fathers are also close relatives.

The jural primacy of patrilineal ties is further evidenced in cases in which a man dies and leaves small children. They are usually taken under the custody of their father's closest male agnate, usually his father or older brother. Thus, if his widow remarries, she will most likely have to leave her children behind. Faced with this, a common but not universal practice is for the widow to marry one of her husband's brothers, if this is convenient. This practice, found among some Jewish and other non-Moslem groups as well, is known as the *levirate.*

From this it is obvious that there is an inherent potential for tension and even conflict between the demands arising from conjugality and the expectations and jural rights based upon lineal ties. Marriage, while serving to reinforce ongoing relationships, can easily lead to contradictory demands on the loyalties and commitments of family members. The expectation everywhere is that sons should cooperate closely with their father, brothers, and close agnates. However, once married and separated from the domicile and direct authority of his father, a man may well find that he is increasingly drawn into the sphere of activities of his wife's family. This is quite understandable if we consider that the wife's sense of identification and desire to be with her natal family is as strong as that of her spouse. Women tend to greatly prize the intimacy and relative informality of their brothers' company, and as a woman bears children she will seek to visit with her family members regularly. When disagreements or disputes arise among close agnates, it is very likely that some will increasingly turn to other relatives for support, including their wife's kin. Affines thus potentially exert a centrifugal pull on a man's loyalty; this is expressed in any number of proverbs and folk sayings to the effect that to allow one's son to "marry out" is to lose him.

The fact that close-cousin marriage is common does not necessarily resolve the potentially conflicting demands on the loyalties of the individual. Indeed, as has been suggested, it may exacerbate conflict under certain circumstances. If a man is married to a close matrilateral or even a distant agnatic relative, already existing social relationships to his wife's kinspeople may be strengthened at the expense of ties to close agnates. Even when a marriage is arranged among close agnates, it is not uncommon for hard feelings to be aroused in the process. There may be other closely related families who also feel that they should have been consulted, or even that their own son or daughter was slighted. In many communities, some of the most enduring intrafamilial disputes involve disagreements over marriage arrangements.

Despite the presence of a strong descent ideology and the importance of patronymic groupings and jural rights stemming from patrilineality, the individual's kin circle of interaction is quite open and flexible. Given the men's ability to shift their alliances to different sets of relatives, families that depend on shared resources such as property or political power have a great concern in regulating marriage. Families whose households rely on wages or individual sources of income are far less concerned with containing marriages within a narrow group. As the economic trend is towards increasingly individuated sources of income, we see a corollary to this in the decreasing importance of close intrafamily marriage and a corresponding increase in the significance of the conjugal bond.

We also see this expressed in the larger patterns of village and neighborhood organization and in settlement patterns. Increasingly, new households are establishing themselves whenever they can gain access to sufficient resources, even when this means moving to distant cities or settling on government-sponsored land-reclamation projects far from their natal families. Settlement patterns for many agricultural regions traditionally reflected the distribution of related patronymic groups. Today this is changing in many areas, as local governments sponsor agriculture projects and village housing, and encourage movement to facilitate rural development.

RESIDENTIAL PATTERNS

As we have noted, the concept of *qaraba*—"nearness" or relatedness—implies spatial proximity. Clearly the prevailing principle is for new households to establish themselves in close proximity to the husband's father, his brothers, or other agnates. This is what Emrys Peters has termed the "patrilineal ethos," a notion of appropriate behavior not necessarily achieved or even inevitably pursued in practice.[19] When feasible, people cluster near the relatives with whom they interact most closely. As we have noted, these need not be, and indeed often are not, one's closest agnates. Still, it used to be unusual for households to establish themselves without close regard for their kin, the primary exceptions being civil servants and professional military personnel who had little choice anyway.

In most big cities today, the severe housing shortage has altered this pattern. In Cairo middle- and working-class households locate whereever they are lucky enough to find accommodations. Entire commu-

[19] Peters, "Aspects of Affinity."

nities are built up around factories. In Baghdad, as in other major cities, the government has sponsored the development of neighborhoods based on occupational groups. For example, officers, engineers, high school teachers, and so on, are entitled to house sites or apartments at special rates. In every country we see the emergence of new neighborhoods that are fairly homogeneous in terms of occupation and class. The increasing mobility of nuclear households and the resulting instability of newer neighborhoods have resulted in a decline in the importance of the social role of the neighbor, or *jar*. The concept of *jar* (*jawar* in Arabic and *komşu* in Turkish) traditionally implied a specific set of mutual expectations and behavior. This included mutual support in times of crisis and regular visiting characterized by informality; in many regards it paralleled the responsibility and obligations one had for close kin.

Although we have just emphasized increasing mobility, most Middle Eastern communities still reflect residential patterns based on kinship. A household may depend on access to land or other resources, which in turn depend on recognition of specific rights and the support of others in the community. When conflict or disagreement arises over property rights, the matter is very often settled within the community on the basis of the support each contender can muster for his claim. One tends to reside when possible near those with whom one is identified politically and socially. Most villages, indeed traditional urban neighborhoods, are politically factionalized, and faction membership is congruent with actual residence in a particular area of the town or village. Even when rural people migrate to the city, they try to establish themselves near already settled kinspeople. Modern housing developments can reflect clear regional, ethnic, or tribal patterns, which in essence is little different from what we see in large American cities, where migrants from different countries tend to cluster together.

The emphasis in this chapter has been on the uses of kinship and kinship networks—groups formed by patrilineal descent, marriage, and patterns of residence. We have tried to introduce the most basic elements of social organization—what might be considered the building blocks of which Middle Eastern society is formed. Our stress throughout has been on the principles that influence group formation and group life; in the subsequent chapter we shift our focus more to the individual in society.

CHAPTER NINE
WOMEN AND THE MORAL ORDER: IDENTITY AND CHANGE

Individuals in any society interact with others, behave in particular ways, and have expectations about others that are shaped by their culture as well as by their individual experiences. Within any given society there exists a structure of shared meanings or common understanding and symbols which relate to sex and gender roles. Even material objects come to signify values and serve both as social markers of identity and as statements about one's self. Our intent in this chapter is to explore some of the values and codes that inform aspects of individual behavior and impart meaning to the lives of people in the Middle East. We undertake this with some trepidation. The systems of meaning and values in any society are difficult to describe analytically. We all too often end up informing ourselves by means of our own values, prejudices, and biases, or by seizing upon a few points of contrasts to describe a complex reality. When you come to the Middle East as an outsider, your senses are assaulted by many images, some of which tempt the unwary to make easy generalizations—for example, about male dominance, sexual segregation, and the codes of honor and modesty. As you stay longer in the region and acquire familiarity with the languages and the cultures, easy stereotypes fall away and the challenge is to fit the wealth of observed variation in attitudes and behavior into any sort of order.

Closely related to this concern with sex and gender roles is the larger question of how model patterns and values, even perhaps personality structures, are played out in a rapidly changing social order today. This is what the literature of an earlier era often termed the cultural and psychological dimensions of "modernization." We take up these issues in turn.

WOMEN AND THE VEIL

A convenient point of entry into the difficult enterprise of understanding sex roles in the Middle East is to focus on one of the most visible and controversial institutions of the area—that of veiling and female seclusion or segregation. Certainly few objects in Middle Eastern society are as charged with meaning and strong feeling for the native and stranger alike as is the veil. The veil and other traditional garments of female covering are often taken as visible symbols of the division of the social world into two contrasting domains—the private sphere of women and the public sphere of men. Social reality is more complex than this simple contrast would indicate.

For some Middle Easterners, the veil represents an elaboration and perpetuation of an archaic patriarchal order, an embarrassing relic of the past, and an impediment to development. For others, it symbolizes a core of an Islamic ethos regarding sexual modesty and morality. For still others, veiling is an explicit political statement, even an expediency with which to reaffirm pride of culture in the face of the assaults of the West. These disparate understandings may well be simultaneously affirmed and acted upon in any one society. During the period leading up to the Iranian Revolution, middle-class college women could often be seen in the streets wearing the shapeless traditional dress, or *chador*. Some may have subscribed to the religious tenets expressed by female veiling. Others were simply making a political statement protesting the corruption of the Western-oriented regime of the Shah and expressing cultural identification with Iranians of all social classes. In Turkey, too, veiling as a political statement was common; however, today it is taken up almost exclusively by those who identify with ultranationalist or religious parties. Women on Egyptian university campuses are increasingly differentiated among themselves by the wearing of various forms of what is called "Moslem garb," or *ziyy islami*. The adoption of so-called Moslem dress reflects the turbulence of an era in which social and political alienation are expressed by some in terms of a militant version of Islam.[1]

The point is that veiling and the values that underly it must be interpreted within a specific social and political context. Although we can treat veiling as a generalized Islamic phenomenon related to basic notions of honor and shame, the behavioral manifestations of it are shaped by local historical, political, and social forces. Veiling and the practice of female segregation are not uniform within any society, and even where seemingly adhered to in extreme form, need not be accurate indications of the social and political roles that women play in that society.

We have argued before and stress again here that Islam is a fundamental source of shared meaning throughout Middle Eastern society. Thus we are not concerned here with distinguishing specifically Islamic from non-Islamic sources for particular values and practices. We are looking at what people say and do and how they interpret their own lives. Whether female seclusion and veiling owe their genesis to a *sura* in the Quran or to pre-Islamic Arabian or even Mediterranean custom is immaterial for our purpose; the practice of veiling and the notions of female modesty and sexual roles developed over centuries and undoubtedly involve many sources. What is important to remember is that these values and practices are perceived everywhere as being part of the Islamic ethos and have to be understood in this context.

SEXUAL MODESTY

Veiling and segregation of women have their base in the code of sexual modesty as spelled out in the Quran, elaborated in the Traditions, and continually reaffirmed in the *khutbas*, or weekly sermons, from one end of the Moslem world to the other. Female sexual modesty consists essentially of virginity before marriage, fidelity after it, and the maintenance of a particular public comportment throughout. If not actually veiled, women must behave as if they were—that is, they are expected to carry themselves in a manner that suggests that they subscribe to the tenets of modesty. A virtuous woman is usually referred to as *mastura*, chaste or covered, as if invisible to the outsider and stranger.

Modesty consists of more than proper dress in public or the avoidance of illicit sexual contact. It is thought of as shaping all aspects

[1] Fadwa el-Guindi, "Veiling *Infitah* with Moslem Ethic: Egypt's Contemporary Islamic Movement," *Social Problems*, 28, no. 4 (April 1981), 467–85,

Yemeni women at a public wash area in a village.

of a woman's behavior. There is a strong connotation of the sacred attached to the image of the virtuous woman. The word *haram,* meaning both sacred and taboo, is also applied to women. *Haram,* which can refer to a religious sanctuary such as the *Ka'aba* enclosure at Mecca, is also used to describe the women of the household and their quarters in it. In some Arabic-speaking communities, the women of the household are collectively referred to as *hareem* of its male head. The different women have different claims to this status. The patriarch's daughters, for example, make up part of his *hareem* because they are sexually taboo, or *haram,* as are his daughters-in-law. His own wife, on the other hand, is part of the *hareem* in that she dwells in the sanctity, or *hurma,* of his household. It is ironic that the term *hareem* has acquired a meaning in English that is virtually the opposite of its original sense.[2]

Islamic and Middle Eastern codes of modesty not only regulate sexuality, but attempt to restrict it to the private domain of the home. Nothing so scandalizes the average Middle Easterner as public displays of sexual affection and intimacy. In a society characterized by male

[2] See Richard T. Antoun, "On the Modesty of Women in Arab Muslim Villages: A Study in Accommodation of Traditions," *American Anthropologist,* 70 (1968), 671–97; and Nadia M. Abu-Zahra, "On The Modesty of Women in Arab Villages: A Reply," *American Anthropologist,* 72 (1970), 1079–88; Antoun's reply pp. 1088–92. Also see J. Pitt-Rivers, *The Fate of the Shechem or the Politics of Sex: Essays in the Anthropology of the Mediterranean,* Cambridge Studies and Papers in Social Anthropology, No. 19 (Cambridge, Eng.: Cambridge University Press, 1977).

vanity and dominance, men do not normally boast of sexual exploits, nor is the "Don Juan image" particularly admired.

The Islamic ethos which reflects a fairly common male view emphasizes the inherent danger to the social order of unconstrained sexuality and the necessity for male control of female sexual behavior. This mirrors a remarkably similar world view found among many Mediterranean people, a view closely tied to the value system of honor and shame.

As we elaborate later, honor and shame are notions about social esteem (or the lack thereof) based on communal perceptions of an individual's social worth. They are ways of expressing "reputations"; those who aspire to public approbation behave in ways which show concern for the opinions of others. Fundamental to this value system is the idea that men are responsible for the behavior of their kinswomen, and that a man's honor is to a large extent predicated on his ability to protect the women of his household. One consequence of this belief is that male-female interaction is seen as being always potentially disruptive inasmuch as it may call into question a man's ability to control or protect the women for whom he is responsible. Romantic love is therefore particularly threatening, challenging as it does male control as well as the family's reputation.

What distinguishes the Middle East is the extent to which concepts of honor and shame are the explicit basis for social action, and the extent to which social institutions reflect this. As Fatima Mernissi put it in her analysis of male-female dynamics in a Moslem society, that of Morocco, "Sexual equality violates Islam's premise, actualized in its laws that heterosexual love is dangerous to Allah's order...the desegregation of the sexes violates Islam's ideology on the woman's position in the social order: the woman should be under the authority of fathers, brothers or husbands."[3]

The code of sexual modesty and the attendant seclusion of women are, of course, important means of effecting male control over female sexuality and reproductive capacity. Further, they assure male control within the household and facilitate the preservation of productive property within the male descent line. Even a woman's claim to her own children is jurally secondary to that of her husband, and in the event of a divorce she loses ultimate rights to her children. Male control is often expressed in Islamic law in terms of a duty to protect women, who are perceived as weak and socially dependent on males, be they fathers, husbands, or sons.

[3] Fatima Mernissi, *Beyond the Veil: Male-Female Dynamics in a Modern Muslim Society* (Cambridge: Schenkman Publishing, 1975), p. xv.

Concepts of human sexuality lie at the core of the ideology which rationalizes male dominance and control. Needless to say, there is no single shared view of sexuality, nor is there any one consistent ideology that cuts across all classes and groups. Besides, there is also a dearth of data on sexual practices in the Middle East. At the level of the Islamic literary tradition, however, there is a great deal of material on sexuality and its role in society. This literature is frequently referred to by politicians and religious conservatives, as well as by the *'ulama*. Since Fatima Mernissi and Abdelwahab Bouhdiba have already reviewed much of this literature, we draw on their work for the following resumé.[4]

In Mernissi's exposition, the Moslem concept of sexuality is close to the Freudian concept of libido, or that of raw instincts as being a source of energy. Sexual instincts themselves are believed to have no connotation of good or evil apart from how they serve a specific social order. The regulation of sexual drive, therefore, becomes a prerequisite in Islam to the maintenance of social order. There is intense popular concern with sexual matters, in particular female sexual behavior; women discuss sexual matters among themselves with an explicitness that may shock the unwary.

Sexual pleasure is recognized and even celebrated in literature and religious writings and is not considered as confined to males. Women are not only thought to experience sexual desire, but they have a right to regular sexual relations with their husbands, and sexual fulfillment is recognized as the woman's due. Citing a tradition of the Prophet, Imam al-Ghazzali, one of the major theologians of Islam, wrote, "The Prophet said, 'No one among you should throw himself on his wife like beasts do. There should be, prior to coitus, a messenger between you and her.' People asked him, 'what sort of messenger?' the Prophet answered, 'kisses and words.'"[5]

It is further believed that women's sexuality is of such power that unless controlled, it can be destructive to the social order. This requires greater powers of reason and self-discipline than women are believed to possess. The code of modesty in deportment and dress and the seclusion of women serve to minimize direct male-female interaction and to limit sexual contact. By the logic of this view of sexuality, one may say that these practices serve to "protect" the women from the consequences of their powerful sexuality and the man from succumbing to it, and that they thus contribute to the social order by minimizing conflict among the men over women. It is significant that a beautiful woman is referred to in

[4] Mernissi, *Beyond the Veil*, and Abdelwahab Bouhdiba, *La Sexualité en Islam*, Sociologie d'Aujourd'hui (Paris: PUF, 1975).

[5] Mernissi, *Beyond the Veil*, p. 10.

Arabic as *fitna*, a word also meaning social chaos, a concept not dissimilar to that of the *femme fatale*.

From this perspective, women appear as passive respondents to male action, yet possessed of a capacity for disruption. Institutions that foster male dominance and sexual segregation become fundamental to the social order. Despite all the local variations in ways of life, class, and ethnicity, the attitudes towards personal status and family are shaped by principles that express this male view of female sexuality to some degree.

HONOR AND SHAME

As previously suggested, beliefs about female sexuality and its control are closely related to the concept of male honor. As Julian Pitt-Rivers, J. G. Peristiany, Pierre Bourdieu, and others have noted, the concept of individual and familial honor is widespread throughout the Mediterranean world.[6] In the Middle East the most commonly used term to refer to honor is *sharaf*, a complex and diffuse notion. "Honor," writes Pitt-Rivers, "is the value of a person in his own eyes, but also in the eyes of his society. It is his estimation of his own worth, his claim to pride, but it is also acknowledgement of that claim, his excellence recognized by society, his right to pride."[7]

The sources of honor, or this "claim to pride," are many, depending on the individual and his or her specific circumstances. These include family origin, piety, prowess, generosity, and even wealth and power when appropriately used. Honor is, in brief, the ability to live up to the ideal expectations of the society. The most fundamental and universal component of a male's honor in the Middle East, however, is closely tied to the sexual behavior of his womenfolk. A man's honor is thus directly related to a woman's sexual behavior and her general reputation. This special aspect of honor is known in Arabic as *'ard* and in Turkish as *namus*.

'Ard, as *sharaf*, can be individual or collective, in that whole families, even lineages or clans, are thought of as possessing a common "fund of honor." *'Ard*, however, as Abu-Zeid describes it for an Egyptian Bedouin community, is exclusively sexual and is affected only by the conduct of women.[8] *Sharaf* can be acquired and augmented through cor-

[6] See the various articles on the subject in *Honour and Shame: The Values of Mediterranean Society*, ed. J. G. Peristiany (Chicago: University of Chicago Press, 1966).

[7] Julian Pitt-Rivers, "Honour and Social Status," in *Honour and Shame*, pp. 19–78.

[8] Ahmed Abu-Zeid, "Honour and Shame among the Bedouins of Egypt," in *Honour and Shame*, pp. 243–60.

rect behavior and great achievement, whereas *'ard* can only be lost through the misconduct of women.[9] Any breach of *'ard,* just as any assault on the man's general sense of pride or *sharaf,* brings shame, *'ar,* or *'ayb.* Of all sources of shame, none is felt as acutely as that occasioned by a breach of the sexual codes, and throughout the Middle East the most potent curses refer to the sexual behavior of one's mother or sister.

Family responses to such breaches as premarital sexual liaisons and marital infidelity vary from ignoring or minimizing the situation to the extreme, probably rare, of killing the woman. At this level, one cannot generalize at all, as both compliance with and response to this general norm depend so much on the individual circumstances and social standing of the people involved, not to mention their general educational and cultural background. Still, such breaches are usually socially disruptive, and the literature of the area is replete with such cases leading to family feuds and other communal violence.[10]

The honor of the males of the household, the political and social status of the family, and the reputation of its women may be interdependent, particularly in small communities. Where a family or a household controls wealth, exercises power through local alliances, and expects public deference, the reputation of its women assumes political significance. Their behavior is read by others as an indication of the overall prestige and influence of their particular social grouping or family. For this reason, women of well-to-to village households or even those of the traditional wealthy urban class may be more constrained in their behavior than women associated with the poorer households.

The values of honor and shame are more than internalized psychological determinants for individual behavior. This value system operates in the social and political arenas to delineate group identity and to conserve social boundaries. In effect, it is often a statement about relative status and power. Throughout the Middle East, powerful families and groups practice close endogamy, keeping their own women within a close circle defined either by kinship or class. The linkage between familial honor and control over women demands this. While influential and powerful families might regularly recruit wives from other less influential groupings, they do not reciprocate. The taking of women is a sign of power and prestige, but the converse connotes relative weakness.

An extreme example of the political significance of male control over women is found in Saudi Arabia today, especially among its ruling class. Even though a strict Wahabi code of behavior applies to all Saudis, the behavior of women of the royal family and the upper elite is a matter of state concern. Women of the royal Saudi clan are absolutely forbid-

[9] Ibid., p. 256.
[10] Antoun, "On the Modesty of Women."

den to marry outside it, and their behavior is strictly monitored. Saudi Arabian women are generally excluded from public life; those few public offices held by women deal exclusively with women. A woman is legally considered the ward of her menfolk; she may not leave the country without her guardian's permission, be it her father, husband, or brother. She may not drive an automobile, swim at a public beach, or be seen without her veil. Strict sexual segregation prevails in banks, schools, and hospitals. Even government buildings may have a special door marked for the *hareem*. Nonetheless, actual behavior is at considerable variance from the spirit of these proscriptions. For example, wealthy Saudi women traveling outside the country dress in the latest and often revealing fashions, and they dine, drive, and smoke in public. In short, they deport themselves much like the other members of this wealthy international class. At home, however, great care is taken to conform publicly to a segregated lifestyle, and great effort is made to minimize the contradiction between private lifestyle and the prevailing Wahabi ideology. This is understandable. The few occasions when public attention has been directed to breaches in the behavioral code have been dealt with a retribution so harsh as to shock non-Saudis including educated Middle Easterners. Surely part of the explanation lies in the fact that the very legitimacy of Saudi rule is to a large measure based on identification with the puritanical Wahabi creed.

The ways in which notions of honor, shame, and sexual modesty are interwoven and how they influence individual behavior are difficult to describe in the abstract. How these values are expressed, manipulated, and interpreted varies greatly in time and place. Even the most commonplace, everyday activities—for example, the normal deference of women to their husbands and brothers, or the way in which children are scolded and shamed in public—may be revealing of these underlying notions. To illustrate how these values inform specific behavior, let us consider the following brief anecdote.

In the summer of 1974, while the Kurdish War was at its height, one of the authors, Rassam, was in northern Iraq doing fieldwork in and around the city of Mosul. Having obtained her research permit, she hired a taxi and its driver, a native of a nearby village. Her notes read:

> *On Tuesday morning, 'Ali my driver came to the hotel to pick me up to go to the village to begin my survey. He was dressed in the traditional garb of his region, which marked him as a rural inhabitant, a* qarawi, *or villager. I sat at the back of the taxi and we drove off. A few miles outside the city we were stopped at a military checkpoint. The soldier ignored my driver and came around to my side of the car, put his head in the window and asked where I was going. I told him that I was on my way to the village down the road,*

upon which he asked to see my identification papers. Discarding my protestations and ignoring my work permit, he made us turn around and go back to the city, claiming that the road was mined and that he could not guarantee my safety.

On the way back, I expressed my frustration to 'Ali and my fears that I wouldn't be able to carry my survey through. Upon some reflection he suggested that we try again in a day or two, but this time he would put on his suit and I should wear the 'abaya *(the black cloak worn by the more traditional women in Iraq). And so we did. When he came back two days later to pick me up, he was dressed like an* effendi *(an urban gentlemen). As we reached the checkpoint, the soldier on duty came around this time to the window of the driver, asked to see his papers, and wanted to know where we were going. Without a direct glance my way, he waived us on, and we continued to the village.*

The first time that Rassam set out, she was the one in the automobile who was socially "visible," being clearly a foreigner to the area and a woman traveling alone in Western dress. Her status was immediately recognized by the soldier, who ignored the driver and asked for her papers. Her dress identified her as a member of urban, educated society, and the fact that she had hired a car marked her as someone of potential significance. As it transpired, the soldier chose not to assume responsibility for the presence of a strange woman in his area.

On the second attempt, by wearing the *'abaya* and sitting next to the driver, Rassam became publicly invisible; the soldier perceived her as "belonging" to the driver. Moreover, the driver, resplendent in his Western-style suit, had acquired both visibility and a certain amount of social standing as well. The confrontation had now shifted to one between the two men, the soldier at the bottom of the military hierarchy and the *effendi* representing the lower bourgeoisie or those who identify with them. The woman in the car ceased to exist in any political sense. This, we might add, is one occasion when being a woman anthropologist in the Middle East was an advantage. A male anthropologist would have had to show his identification papers along with those of the driver, and on being found a stranger to the area would likely have been turned back.

SOCIALIZATION AND SEX ROLES

Individuals are born into families in which they are socialized to fulfill certain role expectations; in the process they come to view themselves as having appropriate social identities. Studies of socialization patterns in

Little girl taking care of her brother in rural Turkey. (Photo courtesy of Ulku Bates)

the Middle East all emphasize early differentiation in the care and handling of boys and girls.[11] A number of generalizations emerge from these studies which, although they ignore class and educational differences, still have considerable utility. One such generalization is that sons are frequently favored at the expense of daughters. Another is that early on children are taught ways of behaving in public that are appropriate to their gender. In many families, a young girl will quickly learn that her brother has first claim to family resources, which include food, living space, spending money, and clothing.

A village mother with no sons is frequently pitied almost as much as a woman incapable of bearing children at all. She may, in fact, be

[11] See, among others, the article by Adbelwahab Bouhdiba, "The Child and Mother in Arab Moslem Society," in *Psychological Dimensions of Near Eastern Studies*, eds. L. C. Brown and Norman Itzkowitz (Princeton: Darwin Press, 1977); also in same volume, Hisham Sharabi and Mukhtar Ani, "Impact of Class and Culture on Social Behavior, the Feudal-Bourgeois Family in Arab Society."

divorced on this account, although a husband might not receive community approbation should he do so. Social recognition and the inherent prestige of having sons is such that women themselves reinforce this value. They treat their sons with favoritism and indulgence. The birth of a son is usually met with celebration, that of a girl is relatively ignored. Later, the son's circumcision is celebrated with joy and public acclamation. No comparable event awaits a young girl; in fact, the onset of menses, which marks her puberty, is considered shameful and is kept secret. In Egypt, where female circumcision is practiced, especially in rural areas, it is not viewed as an event for private or public celebration. It is usually rationalized by those who do it as a means of controlling incipient female sexuality.

Generally speaking, even a wedding is considered a celebration by the boy's family on the occasion of their taking a bride; it is not usually participated in with equal enthusiasm by the bride's family. The bride is expected to be very unhappy on her wedding day, and she often cries at leaving her father's house. A widespread custom associated with village wedding observance in Turkey, Iran, and many Arab countries is a ritual enactment of "bride theft." The bride is taken, almost "kidnapped," by a party from the groom's family amid wails of protest by the women of her household.

In villages and working-class homes, girls are given domestic tasks and responsibilities and are placed under strict supervision from an early age. It is not unusual to see girls of 6 and 7 already charged with the care of younger siblings. In rural homes adolescent girls serve at the command of younger brothers, much as they do later for their husbands and fathers-in-law. When guests are present, women, their daughters, and very young sons stay apart from male gatherings in the house, eating separately in seclusion, while young boys are encouraged to sit quietly with the elder males. Young village boys face fewer demands on their time apart from seasonal work in the fields or in shepherding. Young women, even small girls, are kept busy with domestic chores almost incessantly. The duties of the boys, such as shepherding, accompanying their fathers to market, or apprenticing at some workshop, often take them away from the home, while girls find their activities increasingly restricted to the home as they mature.[12]

Bonds of affection and emotion within the family appear to reflect these specific patterns of socialization. Boys, in particular, become very emotionally attached to their mothers, who remain throughout their lifetimes virtually the only adult women to whom the boys can express

[12] For an ethnographic account of childhood, sex-role socialization, and the family in a Lebanese village, see Judith Williams, *The Youth of Haouch El-Harimi: A Lebanese Village* (Cambridge, Mass.: Harvard University Press, 1968).

affection in public. Fathers are more apt to play authoritarian roles vis-a-vis their sons, with whom they generally maintain distant and formal relationships. Daughters, on the other hand, often enjoy more relaxed and warmer interpersonal relationships with their fathers. As Bouhdiba says, although without reference to a specific community,

> *the mother often plays the role of buffer between the father and his children, and interposes herself if a threat arises. She knows how to impose her mediation, and often manages to get the father to yield. A true joking relationship exists between a mother and her son; even licentious or bawdy remarks or more or less indiscreet allusions to sexual taboos despite the strict customs—are not exceptional.*[13]

In general and within the family context, women are brought up to find their primary ties and ultimate sources of economic security in their relationships to their fathers, brothers, and sons. A repudiated or divorced woman almost always goes back to her father's house or, should he be deceased, then to that of her brother, who is legally charged with her upkeep and protection. Only very recently have some women been in a position to support themselves and maintain independent households after a divorce. They are still few in number. With the exception of some upper-class and professional women, social pressure makes it difficult for a single woman to maintain her own apartment or house. To protect her reputation, she has to live with some mature relative. One evidence of change in this area is the reported increase in women-headed households. Quite apart from the fact that wage-labor migration increasingly takes more and more men out of the home, upon divorce women are more apt to live alone than before. There are a number of reasons for this. One major reason is the reluctance of brothers or other male relatives to take the women and their children into their households. This changing attitude can be explained by the men's own changing circumstances. Not only are families increasingly dependent on wage labor, which encourages nuclear households, but with urbanization, living space itself is at a premium.

Despite sexual segregation and a strong division of labor, the lives of women, by all accounts and appearances, are generally filled with much the same concerns and rewards as are those of their menfolk. The main difference that cuts across all but the elite classes is that women find their lives tied much more closely to family and household. Men, in contrast, lead their lives in the public domain of mosque, market, and

[13] Bouhdiba, "The Child and Mother," p. 133.

workshop. But here, too, care must be taken not to overemphasize the public-private dichotomy. Village men and women work side by side in the fields, and the street vendors of Cairo include many husband-wife teams selling foodstuffs, clothing, and trinkets. Widowed women can be found in any community acting as de facto heads of households. Further, although there is only limited formal scope for women to exercise authority within the household because the house and property are mainly considered to belong to the male head, women do exercise considerable influence in decision making. Not infrequently, they join in public discussion of issues affecting their families and communities, and one should be careful not to mistake overt signs of deference and reticence for powerlessness or passivity. Female seclusion makes it difficult for outsiders to appreciate the reality of a woman's position in Middle Eastern society.

As we mentioned earlier, within the household a strong division of labor prevails. For the majority of women, their lives unfold almost exclusively within the domestic domain, and their social roles are primarily those of mothers, wives, and daughters. Traditionally, they held no public office, nor did they work as artisans, shopkeepers, or craftspeople in the marketplace. When they did work outside the house, they did so with "social invisibility" as domestics and agricultural laborers in situations in which minimal contact with unrelated males was required. Even marketing and the provisioning of the household with goods from the marketplace was predominantly, although never exclusively, male.

Social classes vary greatly with respect to women and wage labor. Traditionally, women from the poor urban and rural strata worked outside the home. Women's labor was thought appropriately utilized only within the household; those households forced to sell the labor of their women to strangers were considered low in status. Upper-class women occupied themselves with their children, leaving household tasks to domestics. A woman's seclusion and avoidance of public and domestic labor were proof of her husband's success and wealth. It would, however, appear that the most oppressive aspects of female seclusion were felt by women of families who could afford to sequester their females but who were not wealthy enough to sustain the lifestyle and diversions of the wealthy.

CHANGING ROLES OF WOMEN

As women have increasingly acquired education and professional skills, this pattern has been changing. Women working today as doctors, teachers, and high-level bureaucrats confer status on their parents and husbands, quite apart from the economic benefits. What is happening to women in general in terms of their social and public roles is, however, complex.

Some scholars have suggested that industrialization has, in fact, degraded the position of women, as they are forced to sell their labor in the harsh surroundings of the factory without immediate benefits in terms of personal independence or control of their own resources.[14] Others see the involvement of women in wage labor as a first step to full emancipation. In fact, there are no easy answers. The transformation of Middle Eastern rural and urban life that we have described throughout this book has had a profound and differently felt impact on men and women, on rural and urban dwellers, and even on different classes and ethnic groups. In some areas agricultural development and mechanization have generated new rural wealth and a class of landowners whose family members of both sexes benefit from access to public education and professional training. At the same time, the consolidation of small farms is driving many peasants off the land and forcing them to lead impoverished lives in urban slums. Here the real social costs may be far greater for women than for their husbands, who at least control what meager resources they live upon.

The one generalization we feel safe in making is that everywhere in the Middle East, including revolutionary Iran, the traditional ideology of male domination and female segregation and confinement is under attack from within and is, in fact, rapidly eroding. Although there are many countervailing forces, often present in the growing influence of Islamic conservatism, men and women are quietly altering many of the traditional patterns of interaction both within and outside the home. Throughout the region, stores, offices, and hotels are increasingly staffed by women. More dramatically, even lower-class women can be seen with their families on public beaches, a phenomenon not seen even a decade ago in most countries of the region.

Whether debated publicly or simply reflected in the practical arrangements and relationships within the household, sex roles are changing rapidly, much as they have in Europe and America over the last decades.[15] It might not be an exaggeration to say that the experiences of European society following the Industrial Revolution are being com-

[14] See Mernissi, *Beyond the Veil;* and the article by Vanessa Maher, "Women and Social Change in Morocco," in *Women in the Muslim World,* Lois Beck and Nikki Keddie (Cambridge, Mass.: Harvard University Press, 1978), pp. 100–123. This volume contains some thirty-three articles dealing with different aspects of women's life in Muslim society. It is also a rich bibliographic source.

[15] The literature on the changing roles and statuses of men and women in the Middle East is extensive. For two good bibliographies on women, see Samira Rafidi Meghdessian, *The Status of the Arab Woman* (Westport: Greenwood Press, 1980) and Ayad Al-Qazzaz, *Women in the Middle East and North Africa: An Annotated Bibliography* (Austin: University of Texas Press, Middle East Monographs, no. 2, 1977).

pressed and acted out in a matter of decades in the Middle East today. We say this keeping in mind the differences that still separate the two cultural areas.,

The sources of change are many, but two stand out: the economic transformations underway in the region and the legal reforms in family laws. As the Middle East has undergone increasing industrialization, commercialization of agriculture, and urbanization, there have been profound social consequences for individual life chances and the family. These consequences are by no means uniform, nor do we have enough empirical studies so far to allow us to form a fully coherent picture. We can, however, point to a number of widespread developments or patterns.

As noted, rural-to-urban migration has involved a residential shift for literally millions of Middle Easterners. The consequences for women are many. Very often males of the household migrate alone, leaving their families behind. As a result, many communities with heavy outmigration see a rise in de facto women-headed households where, in the absence of adult males, the wives take charge in areas formerly the domain of their husbands. International labor migration, which is particularly important in Turkey, Egypt, North Yemen, Jordan, and Lebanon, affects urban households as well as rural ones. One consequence is that women come to live in households with no males of active adult status. In such instances, women often play wider social roles than they had formerly. Another consequence of labor migration is that women themselves may leave the countryside to join their husbands in the cities or abroad. Once there, it is common for them to take wage jobs.

International labor migration by women is limited in numbers but significant in its social impact, especially in Turkey and Egypt. In 1980 there were about 1.6 million Turkish workers and dependents in northwest European countries.[16] Of these, about one-third were women, many of whom arrived in Europe simply as spouses but soon found employment outside the home. Others were recruited directly from Turkey, coming even from remote rural areas. Many parents who would be reluctant to have a single daughter live alone in a Turkish city actively supported their daughters' seeking work abroad; having a family member employed abroad conferred social status, as well as an important source of income.[17] The risk of bringing shame to the household was minimized by the fact that Europe was acknowledged to have a different prevailing code, one in which more independent behavior by females was accepted. This supports our earlier contention that the codes of sexual

[16] Nermin Abadan-Unat, "Turkish Migration to Europe and the Middle East: Its Impact on Social Legislation and Social Structure," unpublished manuscript, n.d., p. 2.

[17] Ayse Kudat, "Structural Change in the Migrant Turkish Family," in *Manpower Mobility Across Cultural Boundaries*, ed. R. E. Krane (Leiden: E. J. Brill, 1975).

modesty and honor and shame have to be understood within particular political and social contexts. Kudat found that Turkish worker families in Germany readily accepted female strangers in their homes and extended hospitality to unrelated single Turkish women, whereas this would rarely be done in the home country.

Despite the increasing involvement of women in wage labor, there still exists the tendency in the Middle East to avoid using female labor, particularly in jobs entailing a great deal of direct male-female contact. Despite the profound changes taking place, the Middle East differs significantly from other non-Moslem developing countries. As Nadia Youssef notes, "In terms of quantitative comparative data, Middle Eastern countries report systematically the lowest female participation rates in economic activities outside of agriculture. In fact, as of 1970 the most developed country of the Middle East (Egypt) had a much lower rate of female employment than did the least developed country in Latin America."[18] Youssef attributes this to the persistent effects of female seclusion and to the women's continuing avoidance of socially stigmatizing public employment.

Migration to the city does not necessarily imply an immediate change in values or style of life for the villager. In fact, a common complaint of long-established urbanites—whether from Istanbul, Cairo, or Aleppo—is that their city has become "ruralized." It is safe to say that the majority of the inhabitants of most cities in the Middle East today are rural-born and maintain their village ties. This means that the distinction between town and country in style of dress, use of public space, and patterns of consumption are becoming blurred.

For women, one paradox is that under certain circumstances a move from a village to a town or city may increase rather than decrease some aspects of sexual seclusion and segregation. Whereas men almost immediately adopt city dress and have more places to go to for entertainment and diversion, women usually retain village dress and may even adopt the veil. Rural women generally move relatively freely and unveiled because their neighbors are also relatives. Once in an urban setting, this is often not the case, and the women may then veil as they find themselves in the presence of strangers. Veiling and the seclusion of women may also indicate an upward social move, as it symbolizes the fact that the family has acquired the means to free women from work outside the home. It is only over the long term that urban residence is reflected in changed attitudes about seclusion and increased emphasis on female education.

In the Middle East, as elsewhere, the increasing participation of women in the commercial and professional sectors will be reflected in

[18] N. Youssef, "Social Structure and the Female Labor Force: The Case of Women Workers in Muslim Middle Eastern Countries," *Demography*, 8, no. 4 (November 1971), 427.

Urban coffee houses are traditional male domains; this one is in Cairo. (Photo courtesy of the United Nations/John Isaac)

increased personal independence and autonomy. Still, wage labor and employment alone do not guarantee personal autonomy. Very often, working-class women, particularly first-generation immigrants, seem to have very limited rights in disposing of the wages they earn. They may turn their wages over to their husbands, or their wages may be little more than sufficient to meet immediate subsistence needs. The vagaries of the labor market, the high rates of unemployment, and the lack of job security do not facilitate a full and rapid transition to full equality for the women.

Concomitant with the economic transformation of the region is the explosion in secular education, a phenomenon more strongly experienced by urban dwellers. It is hard to assess the extent to which the status of women has changed as a result of public education. However, once educated, at even minimal postelementary levels, women become valuable participants in their nation's development.[19] Only Saudi Arabia has been able to afford to exclude their own women from employment. Instead, they hire foreign migrants on temporary work permits. Ironically, this policy opens up opportunities for Lebanese, Palestinian, and Egyptian women who work there as doctors, nurses, and teachers in girls' schools.

[19] See, for example, Kamla Nath, "Education and Employment among Kuwaiti Women," in *Women in the Muslim World*, pp. 172–88.

LEGAL REFORM AND PERSONAL STATUS

As we noted earlier, legal reform is important to the full appreciation of the status of women in the Middle East. This is especially so with regard to the laws that govern personal status and family relations, which deal with the rights, obligations, and constraints of women in their roles as daughters, wives, and mothers. Most of the organized women's movements in the area have as part of their program the demand for basic reform in the laws governing marriage, child custody, inheritance, and divorce. One of the earliest women's movements in the Arab world—that of Egypt—was founded in the early 1920s. It has repeatedly pressed for the abolishment of polygymy and modification of divorce laws. Today, women's organizations exist in all countries of the area, with varying degrees of political influence. Some are completely controlled by the regimes in power and are no more than government agencies. Others are more autonomous and function to stimulate debate on women's issues as they campaign for reform.

Beginning with the promulgation of the Ottoman Law of Family Rights of 1917, virtually every country in the Middle East has introduced reforms into its codes of Personal Status. The Turkish experiment of 1926 is perhaps the most extreme. All *Shari'a* provisions dealing with personal and family matters were abandoned, and they were replaced with the wholesale adoption of the Swiss Civil Code. With the exception of the Peoples' Republic of South Yemen, no country has gone that far, although all countries have enacted new legislation which can be described as reformist in intent. For example, in Lebanon Moslem women can prevent their husbands from marrying a second wife by so stipulating in their marriage contract. In Iraq, a man must request special permission of the court in order to marry a second wife. The traditional unilateral rights of men in divorce have been widely restricted, and women have been accorded more freedom in initiating a divorce, although nowhere do they have parity with males in the matter.[20]

The history of legal reform affecting the status of women in Iran reveals the sensitive nature of these efforts, which are perceived as interference with the intimate and private domain of the family. In 1936 Reza Shah attempted to abolish the veil and to change certain aspects of the marriage laws to give the women more freedom. He was forced to retreat in the face of strong opposition voiced by the religious leaders. Despite this, over the ensuing years Iran saw the gradual promulgation of a

[20] See the different articles on the subject of legal reform, women, and the family by Noel Coulson, Doreen Hinchcliffe, Elizabeth White, Michael Fischer, and Behnaz Pakizegi in *Women in the Muslim World.*

number of reform-minded laws. These were introduced along with widespread efforts over the radio and in the schools to propagate modern values regarding women's roles. The laws of 1965 and 1975, the latter on the eve of the Islamic Revolution, were meant to abolish polygyny and equalize the rights of spouses in divorce. The Revolution has, for now, ended any such efforts, and the current government has, in fact, repealed all secular family legislation. It is hard, however, to imagine that this reversal will be a permanent one, given the massive socioeconomic changes already effected in Iranian society. Educated women in Iran continue to vigorously resist moves by the clergy to restrict their public employment or to impose on them a specific dress code.

Other countries such as Iraq, Syria, and Jordan reveal a pattern of slow and cautious advance in the secularization of family law, and all change introduced meets some opposition from the more conservative sectors of the society.[21] On at least one occasion, for example, the Revolutionary government of Iraq under Qassem formulated legislation to equalize inheritance between males and females, only to withdraw it because of strong opposition. Inheritance laws are an especially difficult issue for reformers to tackle because of their immediate economic impact on men, and because inheritance rules are specifically spelled out in the Quran itself.

Even once enacted, radical legislation of this sort usually runs ahead of actual practice in some sectors of the society. Contrary to law, for example, polygynous households can be found throughout rural Turkey. In fact, recent reports from the government's office of statistics indicate an increase in reported polygynous marriages, although such reporting is bound to be inaccurate as polygyny is illegal. Men may take a first wife using an officially registered marriage and later on another using a religious contract of no legal standing in court. Such second wives may find themselves divorced without any legal claims to compensation or to their children, a tragic topic frequently treated in Turkish literature on rural life.

Reform laws nonetheless have social impact, even when this may not be immediately apparent. Legislation passed two generations ago in Turkey is certainly one reason why Turkish women today enjoy a higher status both within the household and outside it, compared to Arab and Iranian women. When rights are guaranteed by law, even illiterate women may seek and find help in seeing their claims adjudicated. Taking into account the obvious impediments to implementing and enforc-

[21] For the implications of recent legal reforms in the codes of Personal Status in Iraq, see Amal Rassam, "Political Ideology and Cultural Constraints: Women and Legislation in Iraq," in Daisy Dwyer, ed., *The Politics of Law in the Middle East* (New York: Columbia University Press, in press).

ing radical legislation, and given the gap that exists between "laws on books" and social practice and constraints, legal reforms in the area of personal status still have a significance that transcends their symbolic value.

WOMEN AND SELF-IMAGE

There has been very little research published on the self-perceptions and attitudes of women in the Middle East. They are usually treated en masse, as one monolithic and undifferentiated group characterized by passivity and forbearance and isolated in their own separate world. But surely, women, just as men, shape the world in which they live and contribute to its continuity. The traditional system, its values, and practices developed and persisted because women as well as men found rewards within it. As more women today seek rewards and identities that are no longer encompassed within their traditional domain, the value systems of both men and women will change accordingly. Given that the burden of socialization of children rests largely with the women, change in their self-images, attitudes, and expectations will no doubt affect the new generation. The trend emphasized earlier towards smaller family size and nuclear residence will also contribute to the enhancement of the social status of women.

One important dimension of this changing situation yet to be systematically explored is how men and women are psychologically coping with the changes in their customary roles, with their attendant constellation of rights and expectations. One possible source for such an inquiry is through the analysis of contemporary novels and short stories that deal with interpersonal relations among people in different sectors of modern-day Middle East society. A pioneering work is that of Salwa Khammash, *The Arab Woman and the Traditional Society*.[22] Khammash analyzes the fictional characters depicted in leading Arab novels and illustrates through them the dilemmas and frustrations of the young men and women as they attempt to negotiate new bases for their interpersonal relationships that would accommodate the changes in their environment.

Apart from such general observations based on personal impressions and novels, there are few data that allow us to probe meaningfully into these areas. The intense privacy that surrounds sexual, and indeed many other personal and family matters is an impediment to research.

[22] Salwa Khammash, *The Arab Woman and the Traditional Society* (*Al-mara' Al-'arabiya Wal-mujtama' A'-taqlidi Al-mutakhallif*) (Beirut, 1973).

Even though studies of women's new economic roles and participation in public life are beginning to yield a broad picture of rapidly changing behavior, little is known about the attitudes and beliefs that accompany this change. We know little, for example, about how women of different classes and groups perceive their own lives and society, and how they cope with the restrictions imposed on their behavior.

Those few women or men who have expressed themselves on the issue in poetry, essays, or novels tend to take extreme positions. Women like the Lebanese writer Laila Ba'albaki, the Syrian Ghada al-Samman, and the Egyptian feminist Nawal Sa'adawi have all expressed anger, frustration, and impatience with the social position of women.[23] While genuine, this is unlikely to be universally experienced or shared. The few sources that portray ordinary women in the context of their familial roles and daily activities present a less dissonant and oppressive picture than we might imagine. In an often humorous and moving account of women's lives in an Iraqi village, Elizabeth Fernea shows how women entertain each other, offer mutual support and solace, and manage to influence and manipulate their menfolk.[24] At the same time, Fernea reports the presence of undercurrents of resentment and fear by the women of the men who jealously watch over their behavior. Women are not allowed much freedom of movement and have to take great care not to be seen in the company of unrelated males.

A Norwegian anthropologist, Unni Wikan, discusses the life of a group of poor women living in one of the worst slums of Cairo.[25] The image she conveys is that of families, all small, living in extremely crowded conditions in which even finding daily food is a continuous struggle. In this environment, the women assume strong roles in household affairs, and there is little pretense of male authority. The women forge friendship networks among themselves for mutual assistance and entertainment. However, due to intense competition for status as shown by occasional purchases of new clothes or household items, friendships are easily broken and new ones formed. In short, the picture is one of men and women caught up in grinding poverty which renders all kinds of social relationships highly brittle.

In a rare account of the highly segregated society of northern

[23] See the articles in *Middle Eastern Muslim Women Speak*, eds. Elizabeth Warnock Fernea and Basima Quttan Bezirgan (Austin and London: University of Texas Press, 1977). See also Nawal E. Saadawi, *The Hidden Face of Eve: Women in the Arab World* (Boston: Beacon Press, 1982).

[24] Elizabeth Warnock Fernea, *Guests of the Shiek: An Ethnography of an Iraqi Village* (New York: Doubleday, 1965).

[25] Unni Wikan, *Life Among the Poor in Cairo* (New York: Tavistock Publications in association with Methuen, 1980).

Yemen, Carla Makhlouf reports that women there form a coherent sub-society of their own with its own values and codes of behavior, some of which are at variance with those expressed by the men. The suggestion is that while accepting their assigned secondary status, women do not perceive it as arising from inherent "natural" inequality, nor do they take too seriously all the values expressed by the men.[26] Cynthia Nelson suggests that both men and women are involved in a reciprocity of influence toward one another. What should be investigated, she suggests, are the social constructs that facilitate, limit, and govern "negotiations" between the sexes.[27] What, for example, are the sources of power open to women? In what ways can and do women set up alternative courses for men by their own actions? How do women influence men? How is social control exercised? All these are questions that have yet to be answered before we can begin to get a valid and total picture of the world of men and women in the Middle East.

THE INDIVIDUAL IN A CHANGING WORLD

So far we have been discussing some of the values that underlie sex roles as well as the changes in patterns of male-female relations. Certainly, sexually defined roles are among the most important dimensions of an individual's social identity. Social scientists have also been concerned with other, more general, aspects of personality, particularly as these relate to processes of change and economic development. At the heart of this concern is the belief that prevailing traits or attitudinal patterns establish the preconditions and set the limits to economic development. Are certain personality traits correlated with a predisposition towards change or "modernization"? Do people or countries differ in this respect? Can cultures be fairly characterized in terms of modal patterns of personality? Although we do not treat these questions in detail here, these and similar issues underscore a debate that still continues among students of economic development.[28]

One early influential attempt to describe modal personality traits

[26] Carla Makhlouf, *Changing Veils: Women and Modernization in North Yemen* (Austin: University of Texas Press, 1979).

[27] Cynthia Nelson, "Public and Private Politics: Women in the Middle Eastern World," *American Ethnologist* 1, no. 3 (1974), 551–63. Also by same author, "Women and Power in Nomadic Societies of the Middle East," in *The Desert and the Sown*, ed. Cynthia Nelson (Berkeley: University of California, Institute of International Studies Research Series, No. 21, 1973), pp. 43–59.

[28] A recent volume entitled *Islam and Development*, ed. John Esposito (Syracuse: Syracuse University Press, 1980) is directed to the general question of how a cultural system shaped by Islam experiences development.

for the Arabic-speaking world is that put forward by Morroe Berger.[29] In an effort to understand changing Arab society, Berger addresses the question: "What kind of a person is an Arab?" The basic personality profile that emerges is one shaped, according to Berger, by many forces, among which are Bedouin values, Islamic ideology, a long history of poverty and subordination, and particular patterns of child rearing. All these, he adds, contribute to the formation of a modal personality type characterized by a tendency towards self-exaltation, egotism, and exaggerated sensitivity to criticism. There is also the tendency, according to Berger, to conform to group norms and an inability to assert independence as an individual.

This constellation of traits is, in Berger's view, engendered by the demands made on the individual, who is unable to resolve the conflicting claims of self and family:

> *For a long time Arab society has revealed a tension between individual claims and the claims of the group. Incompletely emancipated, the Arab resentment has exploded in the same direction as before: boastfulness, exaggeration of his capabilities, and a tendency to see the slightest skepticism in another as a grave insult....The Arab displays what might be called a negative individualism which is less an affirmation of his own worth than a revolt against the groups that hold him in thrall.*[30]

Among the more appealing characteristics of the "Arab Personality," on the other hand, is an emphasis on hospitality, personal generosity, and politeness, although these are described by Berger as defense mechanisms controlling and masking underlying feelings of aggression and hostility.

Thus, entire populations are characterized as being essentially conservative and politically passive, seeking refuge in Islamic ideas that are far removed from reality. As Berger puts it, he is trying "to show how social history and personal development have continued to produce a society in the Near East in which insecurity, hostility, suspicion and rivalry find their compensation in a strong adherence to a religious ritual, pattern of ingratiation and hospitality, and a limited form of cooperation."[31] The ethnocentrism inherent in such statements hardly needs comment.

In two recent review articles dealing with the characterization of the so-called "Arab and Persian Personalities," F. Moughrabi and Ali

[29] Morroe Berger, *The Arab World Today* (New York: Doubleday, 1964).

[30] Ibid. p. 157.

[31] Ibid. p. 184.

Banuazizi[32] point out the theoretical and methodological weaknesses of the concept of national character and its lack of usefulness in explaining political processes and behavior even within particular countries in the area. For example, although much is made of the psychological consequences of certain patterns of child rearing among the Arabs, virtually all the data are derived from one or two highly localized studies. Most research on national character ignores the fact that these societies are changing rapidly and are highly diversified in terms of class, ethnicity, and the like. A further weakness, as Banuazizi points out, is that even within local communities there are significant differences in such traits as alienation, fatalism, and satisfaction with one's life. Most studies ignore such variability and provide no base line for comparison with other societies when they conclude that an entire population is characterized by some constellation of traits. What does it mean to conclude that the Arabs or the Iranians, for example, are inherently insecure or distrustful? In comparison to whom? And under what circumstances?

Another way of approaching the relationship between individual psychology and cultural change is in terms of attitudes and values, rather than basic personality. Studies of this genre utilize survey techniques in efforts to understand the variability as well as the shared traits. One oft-cited but now dated study is that organized and published by Daniel Lerner in 1958.[33] We note it because it exemplifies an approach that, if not current, is often implicit in many studies. Lerner's approach equates modernization with Westernization, and views individual attitudes and values as significant indicators of change.

Lerner undertook a large-scale project in six countries in the area, in the course of which interviews were conducted with urban and rural dwellers in order to determine the progress each nation had made towards the goals of "modernization." These studies, carried out during the Korean War when American influence in the region was considerable, viewed modernization largely in terms of the American and European historical experience. A "modern" society was characterized as one whose members exhibited personal and intellectual initiative together with spatial mobility and a secular orientation. There was little attempt to justify the assumption that these traits adhering together constitute a "modern" outlook. Certainly there is little evidence today, for example, that individuals who show mobility in residence are neces-

[32] F. Moughrabi, "The Arabic Basic Personality: A Critical Survey of the Literature," *International Journal of Middle Eastern Studies*, 9 (1978), 99–112; Ali Banuazizi, "Iranian National Character: A Critique of Some Western Perspectives," in *Psychological Dimensions of Near Eastern Studies*, Princeton Studies in the Near East (Princeton: Darwin Press, 1973), pp. 210–30.

[33] Daniel Lerner, *The Passing of Traditional Society: Modernizing the Middle East* (Glencoe: Free Press, 1958).

sarily "secularly oriented." The Turkish workers in West Germany are as religious as their stay-at-home counterparts, as are the Iranian migrants to Kuwait and other parts of the Gulf. Among the six countries, Turkey and Lebanon were cited as being the most "modern," a syndrome which also measured their tendency towards political stability and commitment to democracy. Needless to say, these observations have not withstood the test of time. Individuals in each society studied were scored on such traits as "personal impotency," news-item recall, empathy to the modern press, religiosity, and literacy. Using atitudinal indices and literacy scores, people were grouped in categories labeled "modern," "transitional," and "traditional." While emphasizing that tension existed everywhere between town and country, resignation and action, illiteracy and enlightenment, Lerner tried to demonstrate that the tendency everywhere was towards secular modernization. Mass media, Lerner thought, are the key to modernization as they influence individual attitudes. The problem with his study and others like it, we feel, is that they treat individual attitudes and behavioral traits as if they exist in a political and economic vacuum. Even though Lerner refers to political events and discusses problems of poverty and unequal distribution of wealth, his fundamental assumption is that modernization is a state of mind, and that changes in personality and attitudes precede fundamental changes in the socioeconomic system.[34]

Lerner's view was widely shared in the 1950s and early 1960s, when it did appear that Westernization and modernization were virtually equatable, and that the direction for development for the rest of the non-Western world was clearly marked. With hindsight, of course, we see that the "traditionalists" of Iran, conservative as they may be, have risen against their masters and have made a revolution, and that the clergy and Islamic lay militants use modern mass media to spread their message. In Turkey, shortly after Lerner's study, an unparalleled labor migration took place in which over a million workers took up jobs in Germany and other European countries. Both men and women were involved in this displacement, the majority of them rural people of limited education and urban exposure—the very people whom Lerner and others characterized as "traditional."

Nevertheless, it is true that people perceive their world in distinctive ways, and that these perceptions and strongly held values do influence their behavior. One quite successful attempt to collect attitudinal data for purposes of development planning was conducted in 1962 in rural Turkey under the direction of Fredrick Frey.[35] This study

[34] Ibid. pp. 406–7.

[35] Frederick Frey, *Regional Variations in Rural Turkey*, Rural Development Research Project Report, (Cambridge, Mass.: Center for International Studies, MIT, 1966).

was based on the premise that Turkey, like any country, is a land of great regional, economic, and cultural contrasts. As a consequence, the study was designed to compare the attitudes of villagers in different regions, living in different economic and social circumstances, with regard to such factors as willingness to initiate change, respect for authority, religiosity, and use of mass media. No attempt was made to characterize the Turkish peasantry as a population relative to that of other countries, but rather behavioral and attitudinal differences within the country were documented to examine how they might relate to development. Over 6000 interviews were conducted, representing a national sample of all villagers.

As Frey writes in his report:

> *All in all it seems quite apparent that when one moves from some parts of Turkey to other parts, one moves not only to a different topography, different climate and different levels of economic development, but also, often, into a different psychological atmosphere whose consideration would seem to be no less critical for the policy maker than other basic regional characteristics.*[36]

For instance, a number of cognitive indices measuring distrust of outsiders, knowledge of the country, and emphasis on village or familial loyalties over national or provincial ties showed significant cross-regional variability, which often correlated with objective measures of village development.

Some of Frey's findings were surprising as they ran contrary to expectation. One was that the religiosity of villagers in terms of knowledge and ritual observances was much the same (high) for both developed and economically deprived regions. One difference, however, was that more villagers in the better-developed regions excluded more activities from the domain of religious saliency. This is to say that people act on their common faith in different ways, dependent in part on their economic and other circumstances.

Another approach to understanding the relationship between attitudes, behavior, and social change is that of the case study. This is well exemplified by Iliya Harik's work on the political mobilization of the peasants of Shubra, a village in Egypt.[37] Harik shows how changes in attitudes and behavior of the villagers accompanied the basic transformation of the community brought about by the Revolution of 1952.

[36] Ibid. p. 34.

[37] Iliya Harik, *The Political Mobilization of Peasants: A Study of An Egyptian Community* (Bloomington: Indiana University Press, 1974).

Although a certain measure of growth and occupational specialization had already begun before Nasser's Revolution, these were limited in impact and did not affect the majority of ordinary peasants, especially as these were bound to an oligarchic power structure irresponsive to the community. Following the Revolution, the political and administrative structure of the village was reorganized. Power was taken away from the clan heads and absentee landlords, although they did manage to retain considerable influence. Leadership passed initially to farmer-residents and ultimately to poor peasants who possessed the requisite skills. The ruling council of the village and its new cooperatives were governed by residents. Economic relations were also revamped. A third of the village lands were redistributed among some 436 tenant farmers (out of some 1200 households); soon these beneficiaries and other small farmers came to exercise direct political influence in the village. As Harik writes, "Major aspects of the development process such as rationalization of administration, education, higher incomes, mass media exposure, and political participation have been manifested clearly in the recent transformation of Shubra under the Revolution."[38]

What we see from Harik's analysis of the political mobilization of the peasants of Shubra is that, as we might expect from people anywhere, given proper political and economic inducements, they will take full advantage of opportunities to improve the quality of their lives. Although subsequent development of Egypt's rural areas is far from an unmitigated success, the major problems are clearly infrastructural, rather than having to do with inherent psychological constraints or lack of individual initiative.

In this chapter, we have explored some aspects of individual identity, concentrating on the changing position of women, concepts of honor and shame, and family relationships. These topics were singled out because of the light they shed on many other aspects of individual behavior, self-perception, and social expectations. Even so, we have only scratched the surface and we have raised more questions than we have answered.

We have also touched upon a controversial area, that of psychological characterization of Middle Eastern people and society. Research in this area has so far yielded little of explanatory value. More useful than the psychological characterization of whole populations is trying to understand ideational aspects of how people respond to and cope with their changing environment. Like poeple everywhere, Middle Easterners do behave in ways informed by their values, cultural understandings, and all that imparts meaning to their lives. This, of

[38] Ibid. p. 273.

course, is a function of how individuals are reared, educated, and otherwise made members of their society. However, whether it is men or women with whom we are concerned, Egyptian or Turkish peasants, Christian or Moslem Arab merchants, we have to appreciate their behavior and attitudes in terms of specific circumstances and against a backdrop of great variability among individuals.

CHAPTER TEN
LOCAL ORGANIZATION OF POWER: LEADERSHIP, PATRONAGE, AND TRIBALISM

Political scientists who study politics and political processes in the Middle East tend to focus on state-level institutions, national parties, state policy, and international diplomacy. By contrast, anthropologists interested in questions of authority and power tend to limit their focus on individual villages, segments of communities, and tribal entities. The work of anthropologists has mainly been more descriptive than analytical, and it often details patterns of feuding, factional disputes, and different forms of social control. Less common and much needed is a perspective that relates the local patterns to the larger system of which they are a part.[1] This chapter attempts to do this.

Villages and tribes, for example, are almost always treated as discrete political entities with their own loci of authority and power. Richard Antoun, an American anthropologist who has worked on local

[1] Recent attempts to do this include Reidan Gronhaug, *Micro-Macro Relations: Social Organization in Antalya, Southern Turkey* (Skriftserie No. 7, Bergen, Norway: 1974): Robert Canfield, *Faction and Conversion in a Plural Society: Religious Alignments in the Hindu Kush*, Anthropological Papers, Museum of Anthropology, University of Michigan, No. 50, (Ann Arbor: University of Michigan, 1973); also June Starr, *Dispute and Settlement in Rural Turkey* (Leiden: E. J. Brill, 1978).

political processes in Jordan and Iran, has attempted a taxonomy of what he calls "political communities" and has identified each type in terms of distinctive patterns of social control, leadership, and so on.[2] It is a worthy effort but, as Antoun himself admits, a frustrating one. One of the problems, we feel, is that the "boundaries of political communities" in the Middle East, however the community is defined, can rarely be thought of as fixed. The establishment of boundaries is itself as much an outcome as a cause of political action. Also, in any region loci of power are multiple and may shift regularly; thus identification of political communities is contingent on many factors. For some purposes, a group of families or a lineage may constitute the largest effective political community, whereas for others, the effective political community may be the village as a whole or even groups of villages. Sometimes, dispersed tribes or parties organized for political action may form political communities for purposes of social control, loyalty, leadership, and the like. Spatial or territorial boundaries alone do not set limits to political communities; a dispute in the Lebanese parliament between two deputies may easily lead to intervillage feuding in a remote mountain district of Lebanon. The assasination of a newspaper editor in Istanbul may trigger reprisals in a provincial town.

THE ENVIRONMENT OF POLITICAL BEHAVIOR

To understand the political landscape and its often bewildering shifts in loci of power and factional alignments, one has to consider the economic base of Middle Eastern society. There are, we suggest, a number of important constraints that render a highly personalized and flexible system of local-level political organization advantageous. These, in turn, affect the development and stability of supralocal forms of political organization and institutions. Local populations, even individuals, by necessity cultivate and maintain multiple political and social ties with each other which alllow them to cope with prevailing uncertainties of all sorts. One is tempted to say that when it comes to politics in the Middle East, there is more of it there than anywhere else. Let us see why this should be the case and what forms politics takes.

Until this decade, when improved transportation and oil money made food grown abroad accessible, the population of the Middle East was almost completely dependent on local agricultural production. Indeed, even today, the bulk of the population is still largely dependent on

[2] Richard Antoun, in *Rural Politics and Social Change in the Middle East,* eds. Richard Antoun and Iliya Harik (Bloomington: Indiana University Press, 1972).

local food resources. Throughout history, larger urban centers relied on food produced in their immediate hinterlands. With the exception of a few Mediterranean cities, transport was limited to costly overland routes, which precluded haulage of bulk food items such as grain. At the same time, as we described in Chapter 1, the potential for agricultural production is highly variable, with sharp contrasts between productive and marginal lands, between those suitable for crops and those suitable for animal husbandry. Furthermore, within regions members of particular communities were differentiated in terms of their access to critical resources of land and water and in terms of their place in the system of production. A great deal of exchange of food items, labor, and other services exists within village communities and, in particular, between peasants and the towns to which they were inevitably tied. This urban-rural exchange is due to the fact that no farming community is entirely self-sufficient, because tools, clothing, and numerous items of household necessity and consumption are produced by urban craftspeople. Urban populations themselves depend on extraction of food surpluses from the countryside, whether taken in the form of trade, taxes, rent, or tribute.

The result of this variability in productive potential and regional interdependence is that population is distributed unevenly, with concentrations in areas of high potential and low densities elsewhere. Rich lands and prosperous communities often abut poor areas of impoverished populations. Households and local groups often find themselves in competition for scarce resources. Peasants and pastoralists may fight for the same well, just as farmers may contend for control of vital irrigation canals. Communities, neighbors, and even relatives may frequently find themselves in competition, even as they rely on each other for help and exchange.

Environmental uncertainty is another constraint; most farming communities periodically face conditions of drought, erosion, crop and animal epidemics, and the like. This makes understandable the multiplicity of associations and the intensity with which individuals pursue social relations with others both within and outside their own neighborhoods and immediate communities. Observers of social life throughout the Middle East have been struck by the great amount of time and energy individuals spend socializing and politicking.[3] Whether it be

[3] See John Kolars, *Tradition, Season and Change in a Turkish Village*, Department of Geography, Research Paper (Chicago: University of Chicago, 1963); also Brian Beeley, "The Village Coffee House and Changing Social Structure in Rural Turkey," *The Geographical Review*, 4 (1970), 475–93. See S. M. Salim's vivid description of the social and political significance of the tribal guest house in southern Iraq as described in *Marsh Dwellers of the Euphrates Delta* (London: Athlone Press, 1962).

the ubiquitous teahouses of Turkey, Iraq, or Iran, the village guesthouses of Syria, or the urban coffee shops of Egypt, clusters of men meet on an almost daily basis to reaffirm existing ties, to forge new ones, and to keep an eye on activities of others in the community.

THE POLITICS OF SOCIABILITY

Men of influence, be they tribal sheikhs, religious leaders, landlords, or even local party officials, all host a seemingly endless stream of visitors in their homes and offices. In fact, it is not a coincidence that the same word, *majlis,* is used to refer both to national parliaments and to the daily meeting of men in the homes of local notables. Women, too, are active in their own parallel networks and patterns of visiting.[4] Although many decisions are imposed on local communities by outsiders—for example, government agents or landlords—there is a strong emphasis on consensual action throughout. At the level of the village, local camp, or small town neighborhood, many, if not most, decisions affecting public life arise from long discussions, interminable negotiations, and ultimate consensus. Even gossip cannot be thought of as entirely idle. It is often a mechanism for spreading information, influencing decisions, and controlling behavior.

What does all this mean for the individual? And how does this help the household to cope with the problems of livelihood and general security? One obvious result of this intensity of social contact, in the formal men's meetings or in the informal interaction of home visiting, is to maintain a constant flow and exchange of information. People gather and discuss market prices and conditions, hazards along the road, conflicts erupting among neighbors, factors affecting crops and harvests, and, among nomads, even decisions to move camp.

Further, one's very security of ownership of property or rights to use critical resources such as water and pasture may ultimately depend on the ability to maintain an appropriate standing within the community. In emergencies arising from any number of causes, a household will draw on the different ties it has made. Regular acts of sociability serve to affirm current friendships, alliances, and commitments. In a cultural milieu in which even economic transactions are often expressed in a social idiom, it is not surprising that this is so in politics as well. Whom

[4]See the special issue on visiting patterns among women in the Middle East, *Anthropological Quarterly,* 47 (1974). See also the various articles in Lois Beck and Nikki Keddie, eds., *Women in the Muslim World* (Cambridge, Mass.: Harvard University Press, 1978).

one visits, eats with, and is seen with are often public statements of personal and political loyalty.

A common denominator of local politics is the presence of relatively open, ever-shifting constellations of cooperating and contending households or groups. From the swollen villages of the Egyptian delta to the nomadic encampments of Baluchistan, politics is strongly focused on personality and consists largely of changing networks of dyadic ties. These ties are informal "contracts" between two individuals and are based on mutual expectations of loyalty and assistance, rather than on commitment to abstract principles or codes, either rules of kinship of party politics. Even while the substance of politics is carried out in this form, it does not preclude such political structures as tribes, parties, states, or whatever. It does, however, give these structures a distinctive style.[5]

MEDIATION AND PATRONAGE

The personalized approach to politics is exemplified in many ways. One widespread mechanism for social control and integration is that of mediation. *Wasta,* or mediation, as it is commonly referred to in Arabic, describes the role played by intermediaries—whether kinsfolk, religious leaders, or political chiefs—when they intervene to repair personal breaches caused by disputes or to put people in touch with one another for some purpose. Here the *wasta,* or go-between, functions as an all-purpose broker.

Frederick Huxley describes in detail the language and forms of mediation employed in a Lebanese village where *wasta* is used to resolve and contain conflict within the village community. It is also used, as Huxley puts it, to "tap services provided by outside sources of power (e.g. government)."[6] Villagers who want a new road or school would begin their quest by locating someone well known to them who is also in a position to deal with the appropriate administration officials in Beirut. If the individual succeeds in assisting many villagers, he may develop a circle of clients. Ultimately his power would come to rest not simply on

[5] See the article by P. K. Salzman, "Adaptation and Political Organization in Iranian Baluchistan," *Ethnology,* X, no. 4 (1971), 433–44; also Laura Nader, "Communication Between Village and City in the Modern Middle East," *Human Organization,* 24 (1965), 18–24.

[6] Frederick Huxley, *Wasita in a Lebanese Context: Social Exchange among Villagers and Outsiders,* Anthropological Papers, Museum of Anthropology, University of Michigan, No. 64 (Ann Arbor: University of Michigan, 1978).

his office or ability to intervene, but on the number of followers upon whom he can draw.

Even politicians use personalized dyadic ties to build followings and to recruit for their causes. These ties may be considered a form of patron-clientship, with the political figure securing loyalty in exchange for assistance in the form of mediation. In eastern Turkey it is common for politicians and wealthy landlords to enhance their popular followings through the financial sponsorship of the circumcision rituals of boys from poor families. The sponsor assumes certain responsibilities toward the boy, and the parents, in turn, assume a reciprocal obligation, which may be a vote in the next election.[7]

Just as alliance and expectations of mutual loyalty and support are personalized, so are disagreements and disputes. Publicly expressed disagreements of almost any sort are usually considered evidence of personal animosity. When they occur among friends, they might be taken for disloyalty. Friendly disagreement is generally restricted to a small circle of intimates. Consequently, given the disruptive potential of public disagreement, strenuous efforts are made to cover up or disguise the conflicting interests that underlie compromise decisions. What might be regarded as "lying" or "hypocrisy" in some contexts is thus regarded as socially desirable, if not imperative, when it serves to avoid open challenge and possible social rupture.[8] As Richard Antoun puts is, personal interrelationships in most villages are complex, and rarely do people press political challenges in public, as confrontation carries great social and economic risk.

STYLES OF VILLAGE POLITICS

In a comparative study of political and economic organization of two villages in southern Iran, N. Kielstra, a Dutch anthropologist, describes both formal and informal aspects of the political system.[9] Kielstra's observations can be generalized to a great extent. The formal political structure in Iranian villages, as elsewhere, consists of appointed and elected officials. The *katkhoda,* or headman, is appointed

[7] Ayse K. Sertel, "Ritual Kinship in Eastern Turkey," *Anthropological Quarterly,* 44, (1971), 37–50.

[8] Michael Gilsenan, "Lying, Honor and Contradiction," in *Transaction and Meaning: Directions in the Anthropology of Exchange and Symbolic Behavior,* ed. Bruce Kapferer, ASA Studies in Social Anthropology, 1 (Philadelphia: Institute for the Study of Human Issues, 1976), pp. 191–219.

[9] N. Kielstra, *Ecology and Community in Iran* (Amsterdam: University of Amsterdam Press, 1975).

by the government and serves with his elected council of elders, together with a variety of titled religious functionaries, about whom we say more later. The informal world of Iranian village politics expresses itself through a number of loosely structured "conversation groups" focused on particular "opinion leaders." Kielstra describes some of these groups as dominated by young, energetic men. Most groups, however, formed around older men who were influential because of their religious status or wealth. One way or another, all men of the village participate in one or more of these conversation groups, including those government officials who happen to be temporarily assigned to villages—for example, the school teacher or the commander of the gendarmerie. Issues that concern the village as a whole are discussed and debated in these informal groups. Although some decisions such as military conscription and the like are imposed by government authority working through the headmen, most internal village matters represent consensus—or at least compromises acceptable to most sectors.

Within these Iranian villages, as with most communities, interpersonal rivalries and disputes arise over fields and house plots. Even though, as Kielstra describes it, open hostilities do erupt on occasion, public expressions of personal criticism are generally muted. Individuals communicate their disagreements in private meetings and at home in conversations with relatives and close friends. These constitute a restricted and confidential communication network where, unlike the public conversation group, a piece of gossip or a negative opinion can be passed on. Although a recipient of such information is rarely requested to pass on the gossip or personal criticism, it is expected that he or she will ultimately do so. As Kielstra puts it:

> *One can always deny that one has said what is reported that he has said since no other witnesses were present. The expression "he lies"* (dorug miguyad) *does not carry heavy moral overtones in Persian. It is simply a refusal to confirm a piece of information that is passed on in the confidential communication network.*[10]

These networks of "confidential communications" illustrate the use of intermediaries, even at the most basic household level, in communicating with neighbors and others in the village, thus minimizing the risk inherent in face-to-face confrontation. Even requests for minor favors are usually communicated indirectly, and negotiations of importance—for example, arranging a marriage—are always handled through an intermediary.

[10] Ibid., p. 156.

LOCAL-LEVEL LEADERSHIP

Patterns of social control and leadership at the community level take many forms, apart from the informal mechanisms of kinship, conversation groups, and systematic intermediation. In general, patterns of social control reflect the overall structure of authority and power in the local society and thus vary from one community to another. For example, where government presence is strong, even minor disputes may elicit government fines and police intervention. Where government is poorly represented, even major conflicts—for example, over land and water rights—may be resolved by informal mediation, if not by force of arms.

Authority, or the right to exercise legitimate power, emanates from two primary sources: the government or state, on the one hand, and the "moral order," including religion, on the other. Power, in contrast to authority, is the ability to coerce or force compliance and may stem from wealth, strong family, leadership of armed groups (including gangsterism), as well as from the resources of the constituted government.

Even the ubiquitous office of village headman, locally termed *'omda, mukhtar,* or *katkhoda,* displays great variability in the exercise of authority and power. Headmen of a group of adjacent villages may run the gamut from being virtually powerless to possessing great freedom of decision making and coercion. One may be handpicked by a wealthy landlord, another may be the hereditary leader of a powerful lineage; and a third may be simply imposed on a community by the government. All of these bring to their position different sources of power.

Kielstra writes that the elected *katkhoda* of one Iranian village visits relatives in town whenever asked by the government to register conscripts for the army, thus relegating this unpopular task to outsiders. Where households generally have access to sufficient resources, as in small-scale freehold land tenure, the headman may simply represent the consensus of the moment, or he may have assumed a job no one else wanted. Where resources are concentrated in the hands of few, the headman is apt to represent their interests and correspondingly will either exercise considerable power or simply be a facade for others who remain hidden in the background.

AXES OF LOCAL POWER

Although no model exists that adequately explains variability in the organization of local power relations, we can ask what are the major axes around which authority is organized. One obvious source of authority or legitimate power is the government, which may intervene to select headmen and appoint other local officials. Most often, governments use the office of headmen to incorporate selected local leaders into the national administrative hierarchy. Where the government is expanding its control, headmen may be extremely important agents for effecting change and regulating the day-to-day activities of rural people. Revolutionary Iraq and Egypt are cases in point. In both countries, many important reforms in land tenure and political mobilization were implemented through the offices of village headmen.[11]

Alternatively, the government may select headmen from among tribal leaders or already influential families, such as great landlords, in instances where the government chooses to work through established local interests. In a sense this is what happened in Israel when the Israeli government confirmed Arab clan leaders as village headmen. The effect, probably desired by the government, was to strengthen, if not revive, the traditional lineage, or *hamula*, as a significant political structure at the expense of issue-oriented politics.[12] This approach was frequently employed by Ottoman administrators and later by colonial officers. In the midnineteenth century, the Ottomans settled a number of nomadic tribes in Anatolia, Syria, and Iraq by making their traditional leaders headmen and granting them title to land and other favors. The descendants of these families are often prominent among the regional elite of today.

Closely related to state-derived sources of authority is the ability of the government to intervene in local affairs through police action, law courts, and other administrative agencies. In resource-poor areas, the government may simply ignore local politics as long as open rebellion is not evident. Most commonly, however, retaining direct administrative control of rural populations is a continual concern of modern states. Political centralization, in fact, requires direct intervention by the government, and almost everywhere the administration of rural areas is the direct responsibility of ministries of the interior and the national gendarmerie.

While governments recognize headmen as officers of the state, the people themselves may view the government as an intrusion and even an illegitimate source of power. In such circumstances, headmen and other local leaders may find themselves working to keep the government at arm's length, rather than conveying local needs to their official superiors. In Iran, Kielstra reports, the Shah's regime was recognized as

[11] Iliya Harik, *The Political Mobilization of Peasants* (Bloomington: Indiana University Press, 1974); also James Mayfield, *Rural Politics in Nasser's Egypt* (Austin: University of Texas Press, 1971).

[12] Abner Cohen, *Arab Border Villages in Israel* (Manchester, England: Manchester University Press, 1965).

Local gendarme calling on a village headman in Turkey.

a preeminent force within the village, but one without moral authority; it thus lacked any inherent right to rule in the eyes of the populace.

THE MORAL BASIS OF POLITICS

What gives shape and continuity to political behavior in an area is the persistence of a number of what might be called "ideological or moral models," derived either from Islam or from community or tribal sentiment. Newer ideologies are derived from European political philosophy or are adaptations from them. One traditional model for legitimate political behavior already discussed is that derived from the concept of the Islamic community, or the *Umma*. Another model, which we take up more fully in this chapter, is that of tribalism.

One of the important but little studied areas of political life has to do with the relationship between the perceived moral order and the form of the modern state. Even historically there always existed an unresolved tension between the ideals of the Islamic community and the ruling institutions of the moment. A recurring debate over the centuries has been to define the appropriate political vehicle for Islamic ideals. Today this debate takes on renewed urgency from the fact that most of the

An ayatollah exhorting followers at a political rally in Tehran. (Courtesy of the United Nations/John Isaac)

nation-states of the region are recent creations formed by the breakup or transformation of supranational empires and arbitrary colonial boundaries.

Since the late eighteenth century, nationalism in the Middle East has been a potent intellectual force which has closely followed developments in European political philosophy.[13] The course of nationalism as translated into political action was uneven and, until a few years ago, European agents and army officers could readily negotiate alliances with local leaders without the latter being stigmatized as traitors. It was not until World War II that the majority of the people living in the region became citizens of states whose claims to independence rested on nationalist ideologies, as opposed to dynastic and religious precepts.

The result is ongoing disagreement about what constitutes proper

[13] There exists extensive literature on the politics of the modern Middle East, specifically on nationalism and other ideologies. For one particularly elegant analysis of the rise of nationalist thought in the region, see Serif Mardin, *The Genesis of Young Ottoman Thought* (Princeton: Princeton University Press, 1962); see also Leonard Binder, *The Ideological Revolution in the Middle East* (New York: John Wiley, 1964). For a scholarly and readable overview of politics in the area, see James Bill and Carl Leiden, *Politics in the Middle East* (Boston, Toronto: Little, Brown, 1979). The book has an excellent bibliography.

or moral government—a lack of consensus that transcends the differences of the left/right political spectrum with which we are most familiar. Many people find the contradiction between the Islamic moral basis for society and modern or national forms of administration of great concern. The Iranian Revolution, which, momentarily at least, mobilized a united front drawn from the entire society, was possible because of the chasm which had come to separate the government of the Shah from any recognized moral right to rule. The regime's highly touted claim to continuity with pre-Islamic Persian dynasties was viewed by Iranians as a rejection of Iran's Islamic heritage and institutions. Although Iran is an extreme case because of the isolation of its elite class from the populace at large, similar contradictions and tensions are evident in most countries. Turkey, the most secularized country of the region, is undergoing a form of popular religious revitalization with both open and underground agitation for a return to Islamic forms of rule and the reestablishment of the *Shari'a*.

The Arab states of the Gulf are likewise experiencing political unrest, much of which is expressed in the idiom of religion. A large but generally poor Shi'ite minority is increasingly becoming more vocal in its challenge to the existing political order, citing corruption and a betrayal of Islamic principles. While religion is the rallying cry, and political opposition is often organized through religious associations, for example the Shi'a "funeral clubs" of Bahrain, more and more coalitions are emerging based on economic class that encompass various sects.[14]

ISLAM AND LOCAL POLITICS

At the level of the urban neighborhood or village, where people interact on a daily basis, we see almost everywhere the continuing influence of those who articulate that one broadly accepted code, that of Islam. For the most part, the activities of local religious teachers and prayer leaders are centered on providing spiritual guidance. The line between the religious and political spheres of action is not clearly defined, and such leaders are actual or potential political actors. Even though governments use and attempt to regulate the activities of religious leaders, such men remain alternative sources of local authority. In Turkey, following the military takeover of September, 1980, many clergy or religious leaders were detained. In Iraq, Syria, and throughout the region, governments keep the activities of religious personages under close surveillance, par-

[14] Fuad Khuri, *Tribe and State in Bahrain,* Publications of the Center for Middle Eastern Studies, No. 14 (Chicago and London: University of Chicago Press, 1980).

ticularly when they are suspected of membership in such clandestine groups as the Moslem Brothers.[15]

This is not to say that religious personages may not have official positions within the state bureaucracy. In Saudi Arabia and the Gulf states judges, or *qadis*, are drawn from the religious establishment. In Egypt village religious teachers are paid salaries by the state, and in Turkey religious seminaries are run by the government, and their teachers are paid regular salaries as state employees. Religious personages become politically important when they acquire visible influence and a core of followers.

Formal religious learning and sacred descent are two sources of politically significant authority. With a local reputation for charisma, a man may acquire a substantial following. In many respects, this is a religious version of the patron-client relationship. Barth, for instance, identifies two parallel but competing systems of power and authority that coexist in the same community in Swat, a district of West Pakistan. In one, secular leaders whose influence derives from landownership contend for followers; in the other, religious personages or "saints" attract followers and wield considerable power.[16]

Whereas some religious leaders are wealthy in their own right, others rely on donations and gifts from students and disciples. Such men exercise power and influence through their roles in mediation and arbitration. Their moral prestige is thought to put them above family or tribal sentiments, rendering them neutral in disputes. Their neutrality enables them to mediate intervillage disputes, and the fact that they may be part of a larger network of scholars or learned men facilitates intercommunity communication. As the literate in a largely illiterate society, they emerge as opinion leaders. The homes of well-known religious leaders are likely to be visited daily by a cross section of the community. The poor and the helpless come seeking aid in finding jobs or access to land; the distraught come for spiritual solace; and the wealthy validate their social positions by public displays of piety and respect. In a recent study of Iranian clergy, Michael Fisher, documents the way in which clergy trained under famous scholars go on to establish their own networks of influence and prestige.

In certain communities, religious personages may form the only real source of leadership and power. In northern Iraq, for example, among both Christians and Moslem minority communities, local leader-

[15] Richard P. Mitchell, *The Society of the Muslim Brothers*, Middle Eastern Monographs, 9 (London: Oxford University Press, 1969).

[16] Fredrik Barth, *Political Leadership among Swat Pathans*, London School of Economics Monographs on Social Anthropology (London: Athalone Press, 1959). See also Robert Canfield, *Faction and Conversion in a Plural Society*.

ship revolves around the religious personage—be he a priest, *sheikh* or *pir*.[17] Until recently the Shabak, a community of Shi'a sharecroppers near the city of Mosul, relied on their religious leaders, or *pirs*, who were simultaneously heads of mystic orders, judges, teachers, and main mediators between the local population and outsiders. More commonly, however, religious leaders act as counterbalances to secular, tribal, or administrative leaders.

Among the Turkmen of north central Iran, there are two contrasting sources of local leadership and power, each associated with a distinctive form of economic and village organization. Among the Yomut villagers living on the fertile plains around Gombad-i-Kavus, where cotton is the primary crop, much power was (until the Revolution at least) concentrated in the hands of a few tribal notables, or khans. The khans derived their power from extensive land holdings as well as from their ability to draw on their tribal supporters. These powerful leaders, although rarely recognized as headmen, maintained close relations with offices of the Iranian government. The Goklan Turkmen, who occupy the adjacent highland area to the east, differ greatly in terms of local leadership. Lacking great landowning families for the most part, effective leadership is exercised by *akhunds*, men of religious learning and reputation. Though generally of moderate wealth, the *akhund's* primary source of influence rests on the size of his personal following and general reputation. *Akhunds* themselves meet regularly, and their joint pronouncements on matters of community interest carry great weight.

A second source of religious prestige, as we mentioned earlier, derives from sacred descent. Even here there are gradients of prestige, with the greatest accruing to the descendents of 'Ali and the Prophet's daughter Fatima. Lesser lines of holy descent are traced through other relatives of the Prophet or even his companions. This ascribed status may or may not have political saliency, depending on circumstances. Descendants of the Prophet, known as sayyids, command special recognition. Their bodies and property are inviolable, and they are thought to possess the special spiritual quality of *baraka*. Even though many may be undifferentiated from the rest of the members of the community, some manage to employ their status to gain political influence. They may become successful mediators, as they can travel without fear among hostile neighbors. Among the Turkmen of Iran, entire tribes of lineages are thought to be "holy" in this sense. They marry among themselves and they usually reside in separate villages or in their own quarters. Not surprisingly, their communities form a regular line in the

[17] Amal Rassam, "Al-taba'iyya: Power, Patronage and Marginal Groups in Northern Iraq," in *Patrons and Clients in Mediterranean Societies*, eds. E. Gellner and J. Waterbury (London: Duckworth, 1977), pp. 156–66.

border area of two formerly hostile populations, the Yomut and the Goklan.[18]

The power of men of holy descent, whether in southern Arabia, Iraq, or Iran, might be thought of as roughly inversely proportionate to that of the central government. This is not true, however, for the influence of the learned religious leaders, who may find their power strengthened by their ability to mediate local needs within an administrative framework. The literacy of the *'ulama* is a powerful political tool, whereas the claims of holy descent are probably of most significance in circumstances in which descent itself is the primary political idiom—that is, among tribally organized populations.

What has to be borne in mind in looking at local variations in political processes are the ways that people draw upon both moral and material sources of power and translate them into political activity. A religious personage may well acquire wealth, land, and even governmental recognition. A successful secular leader may in time acquire a moral image or even a holy lineage through a "discovered" genealogy. Governments may encourage sayyids to go to troublesome regions to mediate disputes and thereby facilitate state control. Such a policy was followed by the Ottomans in Iraq; a number of "holy families" were invited to come from Medina and to reside in Mosul and other cities. Having come to do good, many did well, becoming large landowners whose lifestyle appears unencumbered by sanctity.

THE POLITICS OF STRATIFICATION

Abdalla Bujra offers an interesting case study of social stratification and political change in a region of the Hadramout in what is today the People's Republic of South Yemen. Bujra's fieldwork was carried out in 1962 to 1963, and his analysis illustrates some of the ways in which politics and religion can be combined, as they so often are in a stratified society.[19] The traditional small-town society in the Hadramout was highly stratified in an almost caste-like manner into a number of endogamous groupings. Distinctions of rank and status were quite elaborately displayed in dress, public ceremonies, and customary patterns of address, visiting, and commensality. The primary mechanism by

[18]The classic monograph on the role of holy lineages in tribal areas is that of Ernest Gellner, *Saints of the Atlas* (Chicago: University of Chicago Press, 1979). Another study on tribal leadership pattern, albeit in Afghanistan, is Schuyler Jones, *Men of Influence in Nuristan* (London and New York: Seminar Press, 1974).

[19]Abdalla Bujra, *The Politics of Stratification: A Study of Political Change in a South Arabian Town* (Oxford University Press, 1971).

which social strata were defined was descent, or the ability to claim ancestry back in an unbroken line to a particular tribal leader or religious notable, or even to the Prophet's family itself.

Towns in the Hadramout, while having a great deal of political autonomy, came under the centralized rule of the sultan's court. At the same time, the countryside was organized tribally, and until 1940 it was virtually outside the reach of the central administration. The division between what Bujra describes as relatively peaceful townspeople and the armed, chronically feuding tribesmen was bridged by the economic interdependence of town and country. Reoccurring droughts and bad harvests necessitated regular shifts of rural populations, keeping the tribes relatively small and poor. Frequently, tribesmen became attached to or allied with urban dwellers as clients. Sometimes the tribesmen worked the land owned by their wealthier urban patrons.

The political life of the town was dominated by a religious group. At the top of the hierarchy were the sayyids, families whose claim to direct descent from the Prophet was recorded in elaborately detailed genealogies. Their social precedence was asserted in virtually every civic domain and acted out in numerous public religious celebrations and feasts in which they played the leading roles. Quite pragmatically, they maintained their power through control of education in the town, a near-monopoly of membership in the public council, and their privileged relationship with the sultan. In short, they used their status to maintain close relations with men of influence throughout southern Arabia, thus perpetuating and reenforcing their local standing.

The second distinct social stratum of Hadramout town society consisted of two groups, again defined by descent. One, inferior only to the sayyids, was a group of learned men who claimed descent from reputed scholars and holy men of the past. Known collectively as *sheikhs,* this group, though called scholars, did not devote itself exclusively to religious scholarship, but tended to own shops and other concerns. Even though the eighty-five families who then comprised this group were divided by conflict among themselves over land rights and other matters, they nevertheless maintained closed ranks when dealing with other groups. Working with the sayyids, the sheikhs exerted considerable political power and could be considered part of the town's oligarchy.

Of approximately similar status to the scholars in the second social stratum were the tribesmen who likewise claimed long descent lines from some tribal hero. Politically, the tribesmen exercised little power but served as clients and allies of the scholars. The third and lowest stratum in the town was that of the *masakin,* people of no recognized descent claims—religious or otherwise. They constituted the majority and served as laborers, craftspeople, and petty shopkeepers.

Today national laws in the People's Republic of South Yemen have abolished all distinctions of rank and inherited status. However, the society has not changed as much as one might expect. As new economic and political opportunities opened up in the 1950s, the upper classes like the sayyids and sheikhs were the ones to take advantage of them. By 1970 these groups had managed to replicate their traditional priviledged roles, even within the newly emerging nation-state. Following Bujra, we might speculate that the current and radical Marxist regime in South Yemen has bypassed this elite and draws on the political loyalty of the large but hitherto excluded mass of Yemeni society.

TRIBES AND TRIBALISM

In the preceding pages, we have used the term *tribe* over and over and in different contexts. We referred to the juxtaposition of tribal and non-tribal populations in South Yemen. We described the political structure of the Turkmen of northern Iran and of the Yörük of southeast Turkey as tribal. We talked about the shift in political organization in southern Iraq as a tribal confederacy was absorbed into a state bureaucracy. At this juncture in our consideration of political processes, it is appropriate to take up the question of tribalism. Our organizing questions are: What makes a tribe? Under what set of circumstances do people use this form of social and political organization, and to what ends? Does tribalism as a form of social identity differ from ethnicity, and under what circumstances does it persist or dissolve? And finally, what can be identified as distinctive to the Middle Eastern version of tribalism today?[20]

In our view, tribalism in the Middle East is best considered as one organizational principle in a dynamic and complex political environment. Other organizational principles include ethnicity, class, and even nationalism. As Fuad Khuri puts it, "Tribalism is not a single phenomenon, an undifferentiated whole, a peripheral social system or simply a stage in the evolution of human civilization."[21] It is rather a per-

[20] The anthropological literature on the concepts of tribe and tribalism is extensive, although the bulk concentrates on Africa. For the central Middle East specifically, we suggest the following: Emmanuel Marx, *The Bedouin of the Negev* (Manchester, Eng.: Manchester University Press, 1967); William Irons, *The Yomut Turkmen: A Study of Social Organization among a Central Asian Turkic-Speaking Population,* Anthropological case study of sedentary tribal population, see Fredrik Barth, *Principles of Social Organization in Southern Kurdistan,* Universitaits Ethnografiske Museum Bulletin, No. 7 (Oslo: Brødrene Jørgensen, 1953).

[21] Fuad Khuri, *Tribe and State,* p. 12. See also E. Marx, "The Tribe as a Unit of Subsistence: Nomadic Pastoralism in the Middle East," in *American Anthropologist,* 79, no. 2 (June 1977), 343–63.

sistent social and political force bringing together people for many different purposes, and doing so in the context of many different, competing, or alternative principles of alignment.

Carleton Coon noted that virtually everywhere one sees the contrast between the "land of the governed" and the "lands of insolence," where the authority of the state finds its limits defined not by national frontiers but by internal opposition.[22] In the past, powerful landlords, charismatic religious leaders, and political dissidents of all kinds have mobilized followers for concerted political action, opposition, or outright rebellion. In Lebanon today, local "big men," or *za'ims*, hold sway and divide the country among them into closely held fiefdoms. Their claim to leadership, apart from raw power, is often based on long-recognized claims of land ownership and patronage. But perhaps the most persistent and pervasive basis for local group recruitment and political action is that of tribalism. Time and time again the tribal idiom has been utilized to express opposition to the state itself, even to topple governments and dynasties. In fact, what most distinguishes the Middle East politically is the persistence of tribalism coexisting with the state.

As we have noted, the state in its various forms has existed for millenia in the area, but so too have alternative and complementary modes of political organization. At times tribalism has been compatible with the state and at times it has been antithetical to the interests of centralized rule. Although the tribe as a form of kinship organization has important social and economic functions, its primary significance in the history of the Middle East has been in the political arena. Ibn Khaldun, a fourteenth-century historian, identifies the notion of *'asabiya*, tribal or group sentiment, as the main driving force underlying historical dynastic change.[23] *'Asabiya* is seen as the sentiment that justifies and legitimizes concerted political action; it is at once a statement of shared genealogical roots and loyalty to the group of people formed on this basis. The sense of *'asabiya*, or identification with a genealogically defined group, can be an important component of an individual's personal identity, as well as the potential basis for group recruitment.

In an earlier chapter on kinship and marriage, we noted that kinfolk of all sorts are drawn upon for important social and economic support. What sets unilineal descent apart from kinship in general is that it

[22] This opposition between core areas effectively governed by the sultan and marginal areas where "anarchic" tribes prevailed has been especially elaborated for Morocco, where an historical distinction is drawn between *bled al makhzan*, the land of governed, and *bled al siba*, the land of violence. For an overview of this, see Ernest Gellner, *Saints of the Atlas*.

[23] Ibn Khaldun's theory of the circulation of ruling elite is based on notions of group solidarity that recall Durkheim's distinction between mechanical and organic solidarities. See Ibn Khaldun, *Philosophy of History* (Chicago: University of Chicago Press, 1957).

defines groups or potential groups vis-a-vis one another; in this way unilineality can become a potent political principle. Unilineal descent is the core concept underlying tribes and tribalism in the Middle East. It creates group boundaries and establishes a framework for relating different groups to each other. The groups formed by this principle may vary greatly in size, ranging from a handful of households to a grouping of many thousands. Members of such a grouping may reside contiguously or they may be dispersed among others. Local groups mobilized by the principle of patrilineal descent—that is, tribes or tribal segments—may control such resources as land, water, and pasture; they may act concertedly to defend a territory or even to promote a special ideology. The actual composition of these groups will almost inevitably include people who are not biologically related or whose claims to common patrilineal descent are dubious or merely putative. What is important, however, is the strength and consistency in employing the claim of common descent to mobilize and rationalize collective behavior.[24]

TRIBAL STRUCTURE

Before we take up the dynamics of tribal politics, we should describe the commonly encountered formal components of tribal society in the Middle East. We use Arabic terms, as they are widely employed, keeping in mind that similar units of organization are distinguished by Persian-, Kurdish-, and Turkish-speaking tribal communities. The basic unit of tribal society, like that of society at large, is the individual household, *al-beit*. This term encompasses the inhabitants together with their dwelling and consists of closely related persons living under the same roof or in the same compound and highly dependent on one another for their livelihoods. The Turkmen description might illustrate the meaning of household or this level of organization; they say the household is simply those people "whose expenses are one."

A number of households whose heads share an immediate patrilineal relationship to an ancestor two to five generations back constitute a second level of organization, which is often called a *fakhd*, or lineage. It is at this point that the tribal idiom defines a group that has potential political significance. Members of the same *fakhd* may share a sense of collective responsibility for property and person, and may

[24] There have been some notable attempts to typologize tribes in the region. See, for example, Richard Tapper, "The Organization of Nomadic Communities in Pastoral Societies of the Middle East," in *Pastoral Production and Society* (Cambridge, Eng.: Cambridge University Press and Maison des Sciences de l'Homme, 1979), and Douglas L. Johnson, *The Nature of Nomadism*, Department of Geography Research Paper # 18. (Chicago: University of Chicago, 1969).

possibly own property in common. The households of such a group are likely to be linked to one another by a history of close interpersonal relations, including ties of marriage, and by the sharing of a common name. They also share a common reputation expressed in the code of honor. Often, but not inevitably, the *fakhds* take the name of their shared ancestor. In theory, both men *and* women remain identified throughout their lifetimes with the lineage into which they are born. Should women marry outside their lineages, they nevertheless consider themselves members of their father's group, and should they be repudiated or divorced, they return to it.

A number of those lineages, again claiming descent from a yet more distant common ancestor, form a larger unit called *'ashira*. Using the same idea of a common ancestor, a group of *'ashiras* may form a tribe, or *qabila*. In practice, both the terms *'ashira* and *qabila* are interchangeable. It is the existence of a number of levels of organization of varying degrees of inclusiveness, all formed by the same principle of descent, that distinguishes tribal society. Moreover, these different levels of organization have different political and economic significance for the individual member of the tribe. For example, in a nomadic pastoral society like the Murra, those tents that camp and move together are most likely to belong to the same *fakhd*. In times of serious conflict with outsiders, more inclusive levels of tribal organization may become operant, bringing together distant relatives who would rarely see each other on a daily basis. In forming and consolidating ever-larger groupings, there is one dominant principle, which is expressed in the Arabic proverb, "I against my brother; I and my brother against my cousin; I and my brother and my cousin against the stranger."

This saying expresses an ideal of how things should be in principle—namely, that closely related individuals and groups should automatically unite, and that claims of blood take precedence over other commitments. In fact, it is recognized everywhere that patterns of cooperation and alliance are not automatic. Lineages and tribes are as apt to be rent by factions and disagreements as are any other communities. The precedence of blood and the idea of solidarity derived from shared patrilineal descent are fundamental principles of organization. They constitute the norm against which behavior is measured and loyalty is judged.

The model for understanding tribal organization was first formulated in 1940 by E. Evans-Pritchard for the Nuer of Sudan and is known as the principle of segmentary lineage. Since then it has been widely used and argued over by anthropologists interested in understanding the political processes of tribally organized peoples in the Middle East and Africa. The segmentary lineage principle tries to answer the important question of how order is maintained in the absence of central power. It does so by positing the principle of complementary opposition, which is illustrated in the proverb we cited earlier. According to this, segments of the tribe or group are expected to ally themselves or coalesce with other groups of the same level according to proximity of descent. It is assumed that all segments or lineages at the same level are roughly the same size and strength and that all individuals in the tribe are equal in status. If this were the case, then it would follow that there would always be a balance of power within the tribe which would dampen disputes and prevent them from ramifying through the group.

The principle of segmentary lineage is an ideal or theoretical construct used by people to describe their own social order. It is not an accurate reflection of behavior. Complementary opposition is likewise an ideal statement expressing the primacy of kin loyalty. As Salzman notes, it is also a statement about the contextual nature of loyalty and alliance. It signifies that alliances among people and groups are impermanent and ever-shifting, as allies are brought together according to the circumstances of the particular situation and especially the relationships that exist between the people involved. In a way peculiar to Middle Eastern tribal politics, professions of loyalty and alliance contain an important caveat: "subject to prior claims."[25]

The assertion of the primacy of patrilineal ties is more often a political metaphor rather than a description of the substance of a particular relationship or alliance. In practice, lineages and tribal segments described in the ethnographic literature contain individuals who are not actually related to the patrilineage or who are related to it by matrilineal ties only. Similarly, as Emrys Peters and others point out, alliances among lineages do not follow the closeness of the genealogical tie. Thus, even though segmentary ideology is a model used by people to order or explain political relations in general, it by no means determines them.[26]

Keeping the preceding in mind, we can say that the segmentary lineage model of society is a shared structural element of Middle Eastern tribes. Another shared aspect of political organization is the vesting in close patrikin of the responsibility for one's social behavior, reputation, and personal security. The code of vengeance and the shared "fund of honor" express and affirm collective responsibility. Although this may appear simply to reflect the lineage model, it actually adds an important complication, one that often sets effective limits to the cohesion of patrilineal groups.

[25] Philip K. Salzman, "Does Complementary Opposition Exist?" *American Anthropologist*, 80, no. 1 (1978) 53–70.

[26] Emrys L. Peters, "Some Structural Aspects of the Feud among the Camel-herding Bedouin of Cyrenaica," *Africa*, 37, no. 3 (July 1967), 261–81.

Whereas common descent is the rationale for forming large named groups, personal security lies in a circle of relatives focused not on an ancestor but on the individual himself or herself. Thus, each individual (apart from his or her siblings) will have a unique cluster of patrikin who bear primary responsibility towards him or her. Elizabeth Bacon has termed such an ego-centered protection or vengeance group "the sliding lineage."[27] Unlike the descent-defined or fixed lineage, the individual's vengeance group is his or hers alone. Close relatives will, of course, have overlapping circles of kin, but some will nevertheless be more closely related to one party than to the other.

The way in which this circle of kin is defined and their rights and obligations vary from tribe to tribe. For example, among the Bedouins of Arabia this group commonly includes all patrikin related to ego by no more than five intervening links of descent. In practice, this includes brothers, uncles, sons, and all first and second patrilateral cousins. This group is often referred to as the *Khamsa*, or Five, and it shares a collective responsibility. Should a man commit a homicide, vengeance could be exacted by killing any male member of his *Khamsa*, although usually vengeance is directed to those most closely related to the murderer. The Yörük of Turkey similarly organize personal security in terms of a cluster of patrikin but do not define this grouping as precisely as do the Bedouin. Vengeance and collective responsibility do, however, extend to first cousins. The Turkmen of Iran term their vengeance grouping "those to whom the blood reaches," and it is defined as those who share a common patrilineal ancestor in seven or less generations.[28]

Because most individuals within a community have unique but overlapping vengeance groups, the potential for conflict to result in serious social disruption is quite high. It must be said, however, that given the overlapping nature of these groups of collective responsibility, people in tribal society are usually very concerned about the behavior of their relatives, and much effort is put into defusing conflict and preventing physical violence. For example, among the Yörük responsible relatives and friends are expected to keep men apart who are known to be on bad terms so as to minimize the chance of escalation of the conflict. Inevitably, however, this personalized system of social control sets effective limits on the size and solidarity that lineage segments can achieve and often is the reason for groups to split and reconstitute themselves as separate entities.

[27] Elizabeth Bacon, *Obok: A Study of Social Structure in Eurasia*, Viking Fund Publication in Anthropology, No. 25 (New York: Wenner-Gren Foundation for Anthropological Research, 1958).

[28] See William Irons, *The Yomut Turkmen: A Study of Social Organization among a Central Asian Turkic-Speaking Population*, Museum of Anthropology, University of Michigan Anthropological Papers No. 58 (Ann Arbor: University of Michigan, 1975), p. 61.

Demographic fluctuations, internal factioning, and infighting or even warfare can easily result in substantial differences in lineage size within a particular tribal system. When the lineage controls land or pasture, smaller declining lineages may simply be absorbed by their more powerful neighbors. When control of resources does not depend on the power of the local segment—for example, when land titles are enforced by the state—small lineages may easily coexist alongside larger ones.[29]

EGALITARIANISM AND TRIBAL SOCIETY

The question of social and economic equality within Middle Eastern tribal society is a complex one. There is great variability among tribal groups in terms of leadership patterns and social and economic differentiation, as well as in fundamental ideologies regarding social equality. For example, the Turkmen, although in many respects fiercely egalitarian, nevertheless distinguish two social categories within each lineage. Individuals are regarded as either *iğ*, free, or *qul*, slave, the latter being those whose ancestry includes someone of non-Turkmen descent. All dress the same, mingle freely, enjoy the same standard of living, and are indistinguishable from one another—with the exception of a slight reluctance for free families to intermarry with slaves, although such marriages do, in fact, occur. Much more important is the pervasive economic stratification based on access to land and other resources in a rapidly developing cash economy. This is quickly creating divergent patterns in standards of living, education, and dress. The tribal idiom of political organization does not preclude the development of socioeconomic distinctions, however much these distinctions may threaten tribal cohesion in the long run.

The case of the classical Bedouins of Arabia is more complicated. Bedouins recognized several distinct categories even within one tribal confederation. Among the camel herders, for example, one lineage was likely to be considered of superior descent and to possess an inherent claim to fill positions of leadership for the tribe as a whole. Also, the tribe may have had a number of lineages that have no descent ties to the dominant one, but might simply be clients—a form of second-class citizenship. At the apex were the noble lineages of the camel-herding tribes of the Peninsula, the so-called pure *bedu* or *asilin*, those of pure descent. The sheep and goat herders, or *shawiya*, enjoyed lesser status. Those *shawiya* engaged mostly in agriculture had the lowest prestige of

[29] For a recent study of the transformation of a Syrian camel-herding tribe, see William Lancaster, *The Ruwala Bedouin Today* (Cambridge, Eng.: Cambridge University Press, 1981).

all. These status distinctions did, to a certain extent, reflect political reality in that camel herders were the most mobile and powerful of the desert tribes. At the bottom of the social scale were the slaves and blacksmiths, who constituted a virtual caste group. Tradition claimed that Allah created the first Bedouin and the blacksmith at the same time. In Arabia, blacksmiths formed a small endogamous group and attached themselves to individual tribes.

The prevalence of inequality and social differentiation among tribal groups might seem to call into question the segmentary lineage theory, which stresses equality of membership of all individuals of comparable genealogical standing. Actually, equality of membership in principle need not entail economic equality. The question of equality or economic stratification in tribal societies cannot be viewed either in the abstract or in isolation from the larger society in which the tribe exists. For example, the Turkmen, whom we described earlier as egalitarian in ideology, today are largely settled in agricultural villages where households may vary greatly in wealth and standards of living. This is increasingly reflected in the actual distribution of political power, as some men exert great influence. Barth describes the nomadic pastoral Basseri of southwest Iran as maintaining a substantial degree of economic equality as a result of both rich and poor households' regularly abandoning nomadism for sedentary pursuits. Of course, the Basseri, when taken to include both its sedentary as well as its nomadic members, is a stratified society, a microcosm of Islamic society at large.

TRIBALISM AND THE STATE

In the preceding pages, we sketched some of the principles and constituent elements of tribal organization in the Middle East. Remaining is the question of the relationship of tribes to centralized rule and the dynamics of tribal formation. When do tribes emerge as important in the national political arena? What is the future of tribalism? What is the relationship of tribalism to ethnicity?

We said earlier that a distinctive feature of tribalism in the Middle East is its integration with state forms of political organization. This is in contrast with tribalism in many parts of sub-Saharan Africa, which makes it risky for anthropologists and political scientists to extrapolate from the experiences of that region. The tenacity of the tribal idiom in the Middle East has to be understood in the context of a long tradition of urbanism and centralized rule.

Tribes are not simply coexistent with institutions of state rule and urban domination; they are an integral part of the total whole. Tribes or tribal confederations were identified throughout history with formation of dynasties and contributed to the military and administrative cadres of a number of states. The importance of the tribal idiom in recruiting, mobilizing, and organizing people has varied historically and according to specific circumstances. But tribes have always been a potent political force. The trend may well be, as we speculate, for tribalism in the modern state to increasingly resemble ethnicity. If this is the case, its primary significance will be less as a source of political mobilization than as a source of individual identity. But this is unclear at this point in time.

Tribalism in the Gulf states, in Arabia, and in Syria and Jordan is very different from the recent Iranian or Turkish experience. In Iran and Turkey, tribalism is politically significant inasmuch as tribal identities overlap with ethnic ones, as with the separatist movements among the Kurds, Baluch, and others. In contrast, in the Gulf and in Arabia tribes are closely integrated into the government apparatus and are represented among the ruling elite.[30] In Jordan, too, tribal elements are prominent among the military. Egypt is the only country in which tribalism is of no political consequence; tribally organized people are few in number and are spatially as well as socially peripheral to centers of power and population.

From time immemorial, control of rural populations posed problems for urban-based rulers and administrators, a problem that persists even today. How can a government control far-flung and highly disparate rural populations, maintain inland trade routes, secure markets, and tax revenues? Even poor areas that do not offer a high potential in terms of revenue may nevertheless be critical in terms of trade routes and the security of adjacent zones of high productivity. Such regions are inherently costly to control and administer. A common response is to lightly garrison such areas; then such administration as exists will depend on local structures of power, among them tribes.

The state was often able to maintain an uneasy and incomplete hegemony through shifting its support among rival tribal leaders and local groups. Throughout their long history, the Ottomans tolerated and even encouraged tribalism, as have other governments in the region. In other areas where logistics and communication problems made tribalism impossible to administer, state control consisted largely of occasional punitive expeditions and of treaties with local leaders. Such was the case in the Arabian Peninsula. Rather than even attempt to collect taxes in such areas as the Kurdish highlands, and the Turkmen and

[30]This is the case in modern Bahrain. See Khuri, *Tribe and State.*

Baluch regions, the state frequently paid off local tribal leaders to keep the peace. In such circumstances, tribes and their leaders were not merely the prevailing forms of political and social organization, but the only ones. This picture has, of course, changed dramatically since the 1920s, as governments now have recourse to air power. The British have earned the dubious distinction of being the first to put down a tribal rebellion by air power; this occurred in Iraq in 1919. However, as recently as 1940, Iraqi tribal leaders were still in the regular pay of the government, and some sat in Parliament until the Revolution of 1958.

For rural populations everywhere, close government administration is an uncertain proposition. Although security of property and relief from banditry are obvious benefits, rapacious tax collectors, absentee landlords, conscription officers, and the like add to the burden of the peasantry and reinforce their suspicion and widespread distrust of governments and their agents. This, in turn, fosters or reinforces a dependency on local loyalties, including those of familialism and tribalism, at the expense of political loyalty to the larger state order. Though it is not uncommon for tribal leaders to work closely with government officers and to emerge as powerful exploitative landlords in their own right, they are at the same time apt to be more responsive to local conditions and demands.[31]

The ideas sketched here can serve as the basis for a general framework for understanding state-tribal relations historically. In a very loose way, they serve as a point of departure for understanding the persistence of tribalism in the Middle East today. The framework rests on several historically recurring processes. One is the common tendency to overtax and exploit the more productive agricultural areas for the benefit of the urban population and the city elite. Another is the tendency of governments to overreach the limits of their actual power in attempting to tax and directly control peripheral areas. This may give rise to local unrest and even organized resistance. Successful local rebellions, however occasioned, frequently encourage neighboring groups to do the same and can quickly create destabilizing economic and political conditions whose effects may be widely felt. Local rebellions interrupt vital communications, occupy the military and other agencies of the state, and generally strain the resources of the government. In order to meet such challenges, governments may well institute policies of harsher taxation, which sometimes foment further unrest. For instance, in his attempt to impose direct central rule throughout Iran, Reza Shah had to engage in full-fledged warfare with a

[31] For a discussion of the changing roles of tribal leaders, see Robert Fernea, *Shaykh and Effendi: Changing Patterns of Authority Among the El Shabana of Southern Iraq* (Cambridge, Mass.: Harvard University Press, 1970). Actually Fernea's description is already out of date following the massive changes effected by the Ba'athist government of Iraq.

number of tribal confederacies in succession. His military success was achieved at considerable expense and left much bitterness, which fatally marked his dynasty's rule. Also, as we noted earlier, in attempting to increase their control over the tribes of the Arabian Peninsula, the Ottomans precipitated the Wahabi movement in the mideighteenth century. A number of similar cases can be cited.

The processes outlined here account for the frequently observed paradox that states rely upon local tribes and their leaders even while they are potentially threatened by them. Indeed, a government may well enlist tribal allies to put down other tribal dissidents, a process which, even if successful, may create a stronger power base for the victorious ally tribe. Tribalism, whether the state is in political ascendency or decline, exists as an alternative form of political organization, one more closely representative of local or regional interests, and thus almost "naturally" competitive with the state form of political organization.

TRIBAL, PERSONAL, AND ETHNIC IDENTITIES

What makes the tribal idiom of political mobilization particularly potent is that it can persist even where its immediate political expression is suppressed or destroyed. The ideology of local group loyalty continues as the pattern for recruitment and organization. Its strength derives from the fact that it calls upon the primordial ties of family and kin, which run far deeper and are more enduring than other claims to loyalty. In short, tribalism is familism writ large.

An important aspect of tribalism quite apart from any particular political manifestation or activity of the tribe is its role as a source of social and group identity. Even where lineages do not own property and cannot mobilize for warfare or other corporate ventures, individuals may nevertheless define their social identities by their tribal affiliations. For example, many urban dwellers—especially in the Gulf states but also in the cities of Iraq, Syria, Jordan, and Iran—continue to identify themselves as members of named tribal groupings. In Baghdad, quarters of the city are named after tribes and tribal segments. Recent migrants seek out fellow tribal members for assistance.

Even where tribal forms of organization have no legal standing, populations may still maintain a strong sense of tribal identity. Associating with a particular tribe and a constituent lineage provides a man or a woman with a local history and quite likely with some specific cultural markers, such as distinctive forms of dress, dialect, and even a special cuisine. Depoliticized tribal groupings may serve as informal organizations for mutual assistance and the pursuit of social activities. Small patronymic groups such as lineages and shallow clans closely resemble tribes in structure, even if not explicitly recognized as part of larger, more encompassing tribal groupings. In many rural areas and small towns, as for example in Turkey and Syria, tribes per se are politically unimportant, but patrilineages or clans of several generations' depth are often important social groupings. Many villages are divided into quarters or wards, each dominated by a core of families belonging to the same lineage. Although they acknowledge a descent relationship among most or all of the lineages in the village, they do not consider this relationship specifically significant, nor do they call themselves a tribe. What is significant here is that the same tribal principles of identity and organization are at work, albeit on a more limited scale—that is, tribalism without tribes.

At this point we have to mention tribalism in conjunction with ethnicity. When tribalism serves primarily as a source of personal social identity, it is virtually indistinguishable from what we would regard as cultural ethnicity in the United States or in any other pluralistic society. Even where party politics, job seeking, and patronage are concerned, the parallels between tribalism and ethnicity are obvious. Both serve as the basis for social communication and help to define a person's network.

Also, tribalism and ethnicity often have historical associations arising from explicit governmental intervention. Often governments have attempted to defuse tribalism as a political force by introducing ethnically alien populations into troublesome regions. For example, in the nineteenth century the Ottomans settled many Turkmen tribes in Kurdish areas, moved Kurds tribes into Armenian areas, and settled Circassians and Crimeans throughout Anatolia, Syria, and Jordan. The expectation was that the interspersing of ethnic groups and tribes would prevent regional movements of rebellion. Tribalism and ethnicity in such heterogeneous areas soon become almost synonymous.

With the rise of modern nation-states in the area following World War I, both tribalism and ethnicity have evolved along convergent paths. Increasingly, tribalism is functioning more as a source of personal identity than as a system of organizing resources and mobilizing people for self-defense. This is understandable, given the fact that modern nation-states attempt to hold a monopoly over modern arms. Nevertheless, in regions where government rule is tenuous, traditional tribal organizations persist, as in northern Iraq, in many places in Iran, and in southern Arabia. Increasingly, however, militant tribalism is associated with the strivings and aspirations of national minorities seeking political rights or even independence. In all of the so-called "nations without states," such as among the Turkmen, Baluch, Kurds, and Azeris, tribalism continues to rally and mobilize individuals as well as to be a source of division within the movements.

CHAPTER ELEVEN
CHALLENGES AND DILEMMAS: THE HUMAN CONDITION IN THE MIDDLE EAST TODAY

What are some of the political and social challenges facing people in the Middle East today? How are the forces of change viewed by the people caught up in them? What, for example, moves Iranian city youth to political action? What underlies the virulence of Turkish civil strife? Why is a movement like the Moslem Brotherhood so threatening to current regimes in the Arab world? What, too, are some of the major ideological debates, the sources of political hopes, of anger, even of rebellion? These questions and similar ones seem remote from the usual course of anthropological inquiry, but difficult and controversial as they are, these are questions which must be addressed if we are to arrive at any understanding of Middle Eastern society today. Even a necessarily limited discussion of these issues is fundamental to any kind of synthesis of such material as we have presented in this book.[1]

We order this discussion in terms of a number of interrelated concepts that serve to organize disparate data and thus clarify some basic issues as we see them. In our conception, three closely related and interdependent processes are at work, underlying the social transforma-

[1] An interesting and well-written essay on this subject is Fuad Ajami, *The Arab Predicament* (New York: Cambridge University Press, 1981).

tions in society and economy we spoke of throughout this book. One process at work in every country is that of political centralization: state building and national integration. Another is the increase in economic specialization and the complexity of the division of labor within society. Third, we see everywhere the rise of new forms of social segmentation, individuation, and differentiation. If these can be said to lie in the sphere of political economy, their repercussions are also felt in the realm of ideology and values. Looking at change in terms of these organizing concepts, we hope, will facilitate positioning Middle Eastern society within a wider comparative context while not jeopardizing that which is unique to it.

POLITICAL CENTRALIZATION

One common denominator the countries of the area share with one another and with many developing countries is the drive for political integration, sometimes called nation building, although this latter term might imply more of an antecedent political void than is warranted in the case of the Middle East. Directly or indirectly, every country in the Middle East has experienced the effects of Western political, commercial, or even military intervention together with the erosion of indigenous state-level political institutions within the past century or two. The specific evolution of independence or revolutionary movements and the rise of particular nation-states is beyond the scope of our discussion.[2] What we, as anthropologists, are concerned with here are the social implications of some of these political developments.

We can begin by noting that is was only after World War I that the present political contours emerged. However, most of the countries achieved de facto sovereignty in their current boundaries only after 1945. Over this relatively short period, essentially one or two generations, the present bureaucracies, administrative institutions, and political apparatus were formed. In some countries, such as the Yemens, the Gulf states, and Oman, the process is even more recent.

Although we cannot entirely ignore the extent to which precolonial and colonial institutions have persisted, nevertheless, since the 1950s the region has experienced a turbulent era of political experimentation. Beginning with the 1952 Revolution in Egypt, every country except Saudi Arabia and the Arab Gulf states has undergone some form of revolution,

[2] A good case study of modern nation building is that of Nadav Safran, *Egypt in Search of Political Community: An Analysis of the Intellectual and Political Evolution of Egypt, 1804–1952* (Cambridge, Mass.: Harvard University Press, 1961).

coup d'état, or civil war. The establishment of the state of Israel in 1948 has greatly contributed to the political instability of the region; other factors internal to Middle Eastern society are also at work. What concerns us is not the particular events and outcomes in each country but the fact that, even against a backdrop of violent conflict everywhere, a succession of regimes have managed to extend state power and administrative control. Whether constituted on the basis of a political party, monarchy, or military dictatorship, the states in the region have progressively extended their authority and assumed new and expanded social and economic functions. As part of this process, governments have taken on increased responsibilities for economic planning, administration of law, social control, and security. These are all domains that were formerly, at least in part, within the purview of religious institutions, family, tribe, or local community. Whether the slogans are those of socialism, nationalism, or even religious heritage, the objective is to consolidate state power in the face of strong centrifugal forces, be they of ethnicity, sectarianism, or regionalism.

These processes of political centralization have not been experienced to the same degree in each country, nor has progress in this direction been a smooth one in any country. In every nation, the extension of state rule has occasioned opposition, sometimes from many quarters representing regional, class, or ethnic interests. The practical manifestations of the drive for political integration include the establishment of direct rule over villages, towns, tribes, and regions, intervention into local patterns of social control, codes of personal status, family law, and education. State control of communications media, national systems of justice, and numerous social services are some of the means utilized to promote the primacy of state authority. In some cases, the processes of state formation and consolidation of power have been abnormally rapid due to the stimulus of extraordinary oil wealth. In other countries, governments or ruling parties have had to pace their efforts more closely in terms of existing economic and social constraints.

With varying degrees of success, every country has instituted birth and marriage registration, identity papers, centrally directed literacy and education programs, military conscription, and regulations concerning property transfers, taxation, and the like. Governmental agencies in some countries—Egypt, for example—go so far as to dictate crops to be grown and to attempt to control small commodity markets. In this fashion, the state strives to regulate even the basic system of rural life.

Increased state-level political control and centralization are evidenced in other domains as well. For example, the educational and other national media are utilized to disseminate the idea of a national culture, emphasizing shared historical experience and promoting identification with the nation-state. In Iraq, state-owned and state-operated television regularly broadcasts plays and serials based on historical epics, the morals of which emphasize national solidarity and loyalty to the "Arab Nation." In so doing, the government seeks to play down, even suppress, potentially divisive ethnic or tribal claims to a separate social identity or cultural heritage. In Iran, publications or broadcasts in languages other than Persian continue to be discouraged; in Turkey, ethnic minorities are euphemized; in Egypt, Christian-Moslem cleavages are glossed over. The Ba'athist regime in Syria continually plays down the fact that its primary source of adherence lies in the 'Alawi community, and in North Yemen the Sunni/Shi'a sectarian differences are similarly minimized, at least in government rhetoric. Rhetoric, of course, does not reflect political realities. In Syria, for example, the country has been on the verge of civil war. Opposition to the near monopoly of power by the 'Alawi-backed regime takes the form of Sunni opposition; the Sunni Moslems of Syria form the majority of the population. Lebanon today is on the verge of disintegration altogether into confessional components.

As we have noted, in every country different regions, ethnic groups, and economic classes experience the impact of political centralization differently. The government's success in exercising its authority is almost inevitably uneven geographically because some regions are strategically easier to control than others. Both facts are basic to appreciating sources of political instability, even open opposition and rebellion. Not only do certain groups, classes, or regions benefit more from national integration than others, but, in fact, the process of integration always seems to entail the suppression of some groups, or at least "the benign neglect" of particular regions of the country. In Turkey, for example, interregional distinctions as measured in living standards and general development have been amplified rather than minimized by political integration, because the balance of political power lies in the more densely populated and industrialized western portions of the country. The same might be said of other countries. From the vantage point of numerous local communities, the costs of national integration are borne inequitably; governments may levy taxes, recruit conscripts, and interfere in local affairs while providing few of the services taken for granted in other regions of the country. States often arouse considerable local opposition as they intervene in matters of land adjudication and reform, family law, or secular education. Opposition may be expressed in overt action when the government appears vulnerable, as exemplified by separatist movements in Turkey, Iran, and Iraq. Even where separatist movements are not underway, the very process of nation building creates what might be thought of as a "fund of resentment," a latent source of civil disobedience, even rebellion.[3] This, of course, is not unique to the Middle East, but is also part of the American and European historical experience, as witnessed in civil disobedience and even wars.

Anthropologists, most notably Clifford Geertz and Lloyd Fallers, have been concerned with the problem of the integration of local communities into new polities, or what Geertz has referred to as "the integrative revolution."[4] This concept refers to the forging of a civil polity in the place of ties of kinship, sect, and local community. Historically, systems of Islamic rule placed few demands on the individual in terms of loyalty to a territorial nation-state per se. Although political allegiance was given to a particular leader or dynasty, membership in the universal Islamic community, or *Umma*, commanded one's final loyalty. Loyalty to the *Umma* need not contradict personal identification with particular tribal, familial, or other localized groupings.

Since their inception in the early nineteenth century, nationalist ideologies mark a radical departure in laying the basis for a concept of "citizenship" based on membership in a territorially defined, secular state. One partial exception today is Saudi Arabia, where the concept of *Umma*, as interpreted by Wahabi doctrine, continues to be the primary legitimizing ideology for the state. But even here, much power is vested in what is essentially a modern bureaucracy and secular civil service.[5] Even after several years in power, Khomeini and his partisans have not really addressed the issue of how to reconcile the concept of *Umma* with that of the nation-state in Iran. While striving for the establishment of a pan-national "Islamic State," the political expression of the *Umma*, they have had, quite paradoxically, to rely on appeals to Iranian nationalism.

The conflict inherent in the often-contradictory claims of religion, ethnicity, kinship, and nationalism are rarely resolved in any country.[6] In Turkey, a country of intense nationalism, a large Kurdish minority continues in an uneasy accommodation with the Turkish majority, while at the same time forces of religious conservatism find sympathetic au-

[3] See, for example, Nikki Keddie, "The Origins of the Religious Radical Alliance in Iran," in *Iran Religion, Politics and Society*, ed. Nikki Keddie (London: Frank Cass, 1980), pp. 52–64.

[4] Clifford Geertz, "The Integrative Revolution: Primordial Sentiments and Civil Politics in the New States," in *Old Societies and New State*, ed. Clifford Geertz (New York: Free Press, 1963), pp. 105–57.

[5] William Ochsenwald, "Saudi Arabia and the Islamic Revival," *International Journal of Middle Eastern Studies*, 13, no. 3 (1981), 271–86.

[6] Hisham Sharabi, "Islam, Democracy, and Socialism in the Arab World," in *The Arab Future: Critical Issues*, ed. Michael Hudson (Washington, D.C.: Center for Contemporary Arab Studies, Georgetown University, 1979), pp. 95–104. The book has several other valuable articles.

diences in all of the many ethnic groups comprising Turkish society. Syria, as noted, is experiencing severe interconfessional strife, even though the government is ostensibly pursuing a secular form of socialism. Compounding the problem of conflicting loyalties is the fact that even secular nationalist ideologies can cut two ways. For example, even though Arab nationalism expresses the unity of all Arab speakers, when espoused by any particular leader or country as a political policy, it is likely to be divisive. This is because the appeal of this ideal is often used to promote narrower interests, for example the primacy of one Arab country of faction over another. One reason why Nasser's appeal to Arab nationalism failed to achieve any lasting political unity or even short-term coordination is because his leadership was inextricably joined to Egypt's national interest.

ECONOMIC DEVELOPMENT

A closely related and parallel set of processes at work are those arising from economic development and change. Political ideology aside, the most universally employed and accepted justification for national integration is economic betterment and equity. National courses of economic transformation are highly varied due to any number of specific political factors, as well as to the availability of resources. A common denominator, we would suggest, is the increasing specialization and segmentation of productive processes within national economies. By this we mean that basic productive processes, including agriculture, involve greater and greater division of labor and differentiation among spheres of activity. Although the rate and extent of industrialization per se are highly uneven, every country is experiencing economic changes whereby even long-established local crafts have become increasingly dependent on producers elsewhere, on long-distance transport, and on the integration of different economic sectors through markets or state agencies. The productive process is increasingly segmented, in that output in even small-scale manufacturing and agriculture is now dependent on material inputs from many widely scattered sources, even from abroad.

Agriculture, even where it is not heavily mechanized, is reliant on chemical fertilizers, hybrid seeds, equipment, and mechanized transport. Of course, where cultivation and processing are mechanized, this dependency on supralocal resources is greatly magnified. In Egypt, for example, a recent economic crisis occurred when the government decreased its subsidy of imported fertilizers, and thus occasioned immediately increased cost of production and higher prices. In Iran, although the country is rich in oil, political dislocations have occasioned a shortage of tractor fuel and other inputs, so that agricultural production has declined.

Of course, the most extreme form of productive specialization is in industrial manufacturing. Modern industrial development can be traced back to World War I, when the heightening of nationalism led to systematic private and public efforts to promote industry[7] in Egypt, Turkey, and Iran. This followed a long period in which indigenous crafts and manufacturing had declined precipitously as a result of European competition and colonial policy.

In every country today priority is given to the industrialized sector, and an overall average of 40 percent of the total investment is devoted to manufacturing and mining. This high rate of investment is often at the expense of comparable attention to agriculture. There are several reasons for this, even apart from military considerations. One is the desire to absorb growing urban populations among whom unemployment is rampant. Another reason is to decrease reliance on imports, although most Middle Eastern industry is itself dependent on technology transfer. A third reason is to satisfy local markets. The major areas of expenditure have been in producing machinery, vehicles, textiles, and petrochemical products, the latter to take advantage of the region's oil and natural gas. Charles Issawi suggests that the Middle East and North Africa could well become leading world centers for production of fertilizers, plastics, and other chemical products. At the moment, however, Middle Eastern industry accounts for only 1 percent of the world's industrial output and lags behind Latin America in this respect. Overall, much of the industrial production is under the direction of the state, although Turkey and Egypt today are trying to promote venture capitalism.

Industrial development and economic specialization of all varieties have had broadly felt social consequences. Whether in the formation of a new proletariat or in the form of massive internal or international movements of people, the social consequences are visible everywhere. Unprecedented changes are underway both in terms of what people do for a living and where they live while they do it.

The growing importance of a relatively new working class is evident in almost every country. In Iran the importance of the new proletariat was attested to by the role oil workers and others played in the 1978 Revolution. At a critical juncture, oil workers organized strikes

[7] For general economic history of the region, see *Studies in the Economic History of the Middle East*, ed. M. A. Cook (London: Oxford University Press; 1970); also *The Economic History of the Middle East: 1800–1914*, ed. Charles Issawi (Chicago: University of Chicago Press, Midway Reprint 1975).

that hastened the collapse of the Shah's regime. Later, worker organizations took control of many industries. Even though at this writing there is little active mobilization of industrial workers in Iran, it is possible, perhaps likely, that organized labor will emerge as a force on the political scene in its own right. In Turkey, unions constitute an important force in the national political arena, and the salaries and benefits of industrial workers have outstripped those paid to even midlevel state functionaries. This is the outcome of several decades of labor organization and activism.

Economic investment is unevenly distributed in every country. Nationally directed investments of all sorts are usually concentrated in particular economic sectors and geographic regions. The result of this on a national level resembles what, on a global scale, is sometimes referred to as "the development of underdevelopment." Put simply, some regions benefit at the expense of others. This is because industrial and commercial activities tend to be concentrated in one or two zones, often near the capital or major ports, while other regions languish. Raw materials may be extracted in one region and industrially processed elsewhere where a better infrastructure is in place. The result is that employment opportunities and public services are unevenly distributed. Different regions experience different rates of economic development and population growth. The interregional disparities and resulting inequities in such services as education, health, and consumer goods, not to mention standards of living in general, stimulate interregional migration. As we have already noted, there is a high rate of rural-urban movement throughout. Governments aware of the political implications of this imbalance sometimes attempt to decentralize industrial development, usually without much success.

One has only to witness the massive disparity between eastern and western Turkey to understand why there has been a massive exodus from the poorer provinces. In Iraq the migration to Baghdad and other areas where construction and factory work are available has left some districts with a shortage of agricultural labor that is only partially alleviated by the government's inviting Egyptian farmers to settle in some places. In many respects, the most serious population displacements occur in the agricultural sector, where mechanization alters previous patterns of labor deployment and access to land. Also, many marginal agricultural areas in which agricultural intensification is not feasible lose population as people are attracted to jobs elsewhere. This may further amplify the marginality and relative underdevelopment of such places.

The economic transformations upon which we have touched establish the environment in which unprecedented movement of people is taking place, both within national boundaries and internationally. The metropolitan areas of every country in the Middle East have by now experienced several decades of rapid growth, most in excess of 2.5 percent per annum. Although high birth rates and declining infant mortality underlie this growth, most of the expansion of urban population per se is due to rural migration. How this influx of people has been socially and economically accommodated has to be considered on a case-by-case basis. Apart from the oil-rich states, there is substantial urban unemployment. A growing source of social and political tension is that as unskilled rural migrants swell the potential work force, the relative value of unskilled, illiterate labor is declining. The result is that there is a significant segment of the urban population that will remain economically and socially marginal because it lacks the skills to participate fully in the emerging economic order.

Interestingly, despite the rapid growth of the industrial sector, a number of countries have been exporting skilled workers and professionals. Turkey, Lebanon, and Egypt are foremost in this regard. Turkish and Egyptian doctors, engineers, technicians, and educators are found in significant numbers in Libya, Oman, Kuwait, and Saudi Arabia. This is in addition to the nearly 1.5 million Turkish workers found in Europe. Egyptian school teachers, nurses, and college instructors form an important segment of the health and education establishments throughout the Arab world. Of course, the case of the Palestinian people is a special one, but they too represent a population heavily caught up in international labor migration. Palestinians constitute a highly educated and skilled population as a whole and are found widely dispersed throughout the Arab world filling positions ranging from the skilled construction worker to the highest ministerial ranks. North Yemen has almost a million workers in Saudi Arabia and the Gulf states, most of whom are unskilled. Remittances are an important, even major, source of hard currency for these labor-exporting countries. In North Yemen, food production has become seriously neglected and has been replaced by imported food made possible by the inflow of cash. Turkey, largely self-sufficient in terms of food, is nevertheless highly dependent on worker remittances in determining its balance of payment. In all countries this new source of income has exacerbated rates of inflation as it accelerates a move towards a so-called consumer economy.

The long-term costs and consequences of the export of skilled workers and professionals is unclear, but it cannot be ignored. Egypt, for example, regularly loses its best trained and most highly educated in what amounts to a "brain drain." On the other hand, in Turkey, many workers and technicians have returned after long periods of European employment with capital and skills that have been used to further expand Turkey's industrial and service sectors. The increase in economic interdependency among countries of the Middle East is directly analogous to the heightened integration of regions within particular nation-states. This process of economic interdependency is quietly taking place despite the vagaries of local and international political alignments. Egypt, for example, although at this time of writing is politically isolated in the Arab world, nevertheless remains an important part of the regional economic scene. Turkey, though not closely allied to any other Middle Eastern country, has recently decided to cooperate more fully economically with both the Arab bloc and Iran. Much of Turkish industrial export today is directed eastward towards Egypt, Iraq, and Iran.

SOCIAL AND CULTURAL DIMENSIONS OF CHANGE

The changing economic and political environment has its social, even moral, concomitants. We will try to synthesize what we take to be some of the major manifestations, looking at certain aspects of the social structure and at what might be called the moral basis of the social order. New modes and relations of production have given rise to groups and classes of people constituted along relatively new axes. New elites and new socioeconomic classes or groupings are in the making, crosscutting preexisting social strata. The extent to which processes of class formation in the Middle East are similar or parallel to the European experience is highly debatable. What is incontrovertible is that new elites, indeed new social orders, based on new lines of social differentiation, are emerging.

Through the 1950s, sociologists and other observers frequently wrote about the emergence of the middle class and the muting or even disappearance of traditional segmentation along ethnic, tribal, or sectarian lines. Although we know now that the political saliency of these latter sources of social identity remains and in some cases may even be amplified, it is true that new social formations have evolved. The term *middle class* with its European connotations may be inappropriate, as no comparable social grouping in any Middle Eastern country dominates both the state apparatus and commerce. Still, there is reason to think that these are new and distinctive social formations in the Middle East, sharing among themselves values and material aspirations that set them apart from traditional society. Further, these values and aspirations also closely resemble those of the European middle class, with an emphasis on individuated family structures, secular education, and a strong identification with profession or career.

One of the many new universities in the Arab world; this one in Baghdad opened in 1963.

As Manfred Halpern has written:

> *In the Middle East . . . the new middle class springs largely . . . from groups that had not hitherto been important, and hence had more reason and less deadweight to take advantage of new knowledge and skills . . . The new middle class itself does not define or crystallize its character from the very outset, but only as its various strata come to intervene in the process of modernization and assume additional roles in it. It originates in the intellectual and social transformation of Middle Eastern society . . . as a secularized action group oriented towards governmental power.*[8]

Many of the economic and social roles of the western European middle class are the domain of the state in the Middle East as managed by civil servants, technocrats, or military officers. A more direct analogy is, therefore, to be found in the eastern European countries, where the public rather than the private sector dominates.

The literature on class formation and sources of the new elite is highly polemical, perhaps because there are so few empirical data

[8] Manfred Halpern, *The Politics of Social Change in the Middle East and North Africa* (Princeton: Princeton University Press, 1963).

available. Moreover, the situation varies from one country to the other; such factors as the basis for elite recruitment, the relative power and prestige of the vying factions, and even the extent to which a constituted elite or power group can perpetuate itself differ greatly.

In Egypt, Nasser's experiment in state socialism was virtually abandoned by Sadat, but even Sadat's ability to reshape the power structure was very limited. Many institutions and social forces brought to the fore under Nasser persisted. The Nasser Revolution almost succeeded in disenfranchising the established landowning and commercial elite and replacing them in many sectors of the economy and government with a new technocratic class of very different social origins. This new class of officers and educated civil servants came for the most part from families of modest means. As the public sector grew, this group achieved great social prominence. With Sadat, the situation rapidly altered and the prominence and influence of this new elite was reduced but not eliminated, and the old groups reasserted themselves. As a consequence of Sadat's policy, which opened the country to free enterprise and foreign investment, an entrepreneurial class gained power and prominence. Today, following Sadat's death, the government is attempting to curtail some of the excesses of Sadat's economic policy.

This new entrepreneurial class, whether independently financed or acting on behalf of foreign investors, has by its conspicuous consumption and its flaunting of traditional values already generated strong opposition from within the traditional lower classes and from among the salaried middle class. The same happened in Iran where, under the Shah, a professionalized, secularly educated administrative cadre and a freewheeling entrepreneurial group replaced the formerly dominant *bazaris*, or merchants, clergy, and landowning elite.

The highly variable and frequently changing role of political parties should give pause to anyone considering a simple description of the region in terms of any one basic pattern of class formation or elite recruitment. One generalization that seems to hold true, however, is that rule by the elite of the moment, however it is recruited and organized, is strongly authoritarian everywhere. Only Turkey has experienced significant periods of democratic politics with independent and contending parties. Apart from this, whether the legitimizing ideology is Arab socialism, Islamic justice, or nationalism, the fact remains that power inevitably rests within a fairly narrow segment, with few avenues open for popular participation in the political process.

Although in any given country the cast of political actors active at the moment can be readily identified, those waiting in the wings are understandably harder to identify, as events in Iran have shown. Even where a seemingly stable arrangement emerges among contending sources of power, as was the case in pre-1975 Lebanon, one has to be wary about predicting its endurance. In a 1972 article, Iliya Harik (among others) discussed Lebanon as a successful case of interethnic accommodation, perhaps even a model for other countries. Sadly, subsequent events not only proved Harik wrong, but clearly underlined the key role of outside interventions in the political development of the region.[9]

The formation of new social orders and the changes in patterns of intergroup relations in every Middle Eastern country are accompanied by fresh questioning of what constitutes the proper social contract or the moral basis for society. At the personal level, this is expressed in questions about the individual's rights and responsibilities regarding family, ethnic community, and nation. Even one's primary sources of social identity and the roles appropriate to one's gender come into question and may be resolved in very different ways by individuals in different segments of the society. Much has been made in the literature concerning problems arising from the increased emphasis on personal autonomy for both men and women. This is related to a change in the basic devision of labor within society and occasions new forms of interpersonal relationships of all sorts.

Finally, and of paramount concern to Middle Eastern intellectuals and political activists alike, there is the yet-unresolved question of what is the proper role of Islam in public life. 'Ali Dessouki, an Egyptian political scientist, suggests that the historical experience of the Middle East is distinct from that of Europe. Middle Eastern society must simultaneously cope with a number of major problems, foremost among which is the legacy of defeat at the hands of Israel. In addition, each country faces serious problems arising from industrialization and the unresolved quest for social and economic justice. As a consequence, Dessouki writes, political ideologies tend to stand in polar opposition to one another, and to be what he terms "holistic." They offer all-inclusive remedies for social ills and total explanations for the general state of society. Both Arab secular socialism and Islamic militancy, the two most potent ideologies of the day, exemplify this. Coming from different intellectual sources, each calls for a fundamental restructuring of society—that is, they call for total revolution.[10]

Scholars concerned with the political culture of the area point to the 1948 defeat of the Arab armies and the formation of the State of

[9] Iliya Harik, "The Ethnic Revolution and Political Integration in the Middle East," *International Journal of Middle East Studies*, 3 (1972), 303–23.

[10] Ali Dessouki, "Arab Intellectuals and Al-Nakba: The Search for Fundamentalism," in *Arab Society in Transition*, eds. Saad Eddin Ibrahim and N. Hopkins (Cairo: American University Press, 1979), pp. 418–39.

Israel as the direct stimulus for the most comprehensive and introspective assessments of the ills plaguing contemporary Middle Eastern society. The secular response, most forcibly and influentially stated by S. al-'Azim and N. Bitar, sees the defeats of 1948 and 1967 as the consequences of the failure of Arab society to commit itself to secularism, scientific education, and the eradication of class distinctions.[11] The Islamic response, which has antecedents going back to the late nineteenth century and the struggle against Western encroachment, views the defeat by Israel as the result of the decline of religion in daily life and politics. "Islam and Islam alone as a religion and a civilization is the only condition for our existence and survival as a nation and a culture," writes D. S. al-Munajjid, a leading Arab intellectual.[12] In this view, it is Islam that gave the Arabs their glory, since lost, and it is the adherence to religion that gives Israel its solidarity and its will to prevail.

Khomeini's key role in the Iranian Revolution has more than ever drawn the attention of the world to what some call "the resurgence of Islam" in national politics, and more generally to its role as the focal point in the revival of an interest in the Islamic heritage as a basis for Middle Eastern society and culture. There is, of course, no dearth of new books and articles on this subject. What is perhaps more remarkable is that the saliency of Islam in the social structure of the area has been ignored so often. Western analysts of Middle Eastern social change tended to assume that Islam as a social force was inevitably in retreat in the face of nationalism and modern socioeconomic developments. Also, by concentrating on the ideational dimension of Islam, they failed to appreciate its psychological importance to the individual and its capacity for group mobilization.

As we frequently have noted in this book, religion has been a politically potent force both in the nineteenth-century resistance movements to European expansionism and colonialism and later in the independence movements. Today much of Islam's dynamism is internally directed, in particular towards those forces and institutions perceived to be responsible for the inequities and corruption that plague society. The resurgence of the call to an Islamic morality and social order is closely related to popular perceptions of national and cultural humiliations suffered at the hands of foreigners. This was the case in Egypt following the 1967 defeat, which marks an important point in the history of Islamic movements in that country.[13] In Iran, too, the power of

[11] Ibid., p. 444.

[12] D. S. Al-Munajjid, in Dessouki, Ibid., p. 443.

[13] One of the best articles to appear on the current Islamic movement in Egypt is Saad Eddin Ibrahim, "Anatomy of Egypt's Militant Islamic Groups," *International Journal of Middle East Studies*, 12 (1980), pp. 423–53.

the Shi'a clergy appears closely linked to the increased prominence of the American presence in the 1970s, and in Turkey the rise of Islamic conservatism closely paralleled the failure of Turkey's American- and European-focused economic and political policy. In short, political, economic, or social crises rapidly evoke responses that are expressed in the idiom of religion. This is inevitable because any questioning of political authority, women's roles, family structure, or relations with the outside world may be viewed as impinging upon Divine Law and the moral basis for society which Islam purports to be.[14] However, whether expressed in the idiom of Islam, Arab socialism, or Marxism, there is little doubt that a process of reassessment and searching is underway.

[14] Mohammed Al-Nowaihi, "Problems of Cultural Authenticity and Modernization in Islam," in *Arab Society in Transition*, pp. 581–91.

INDEX

Abdel-Fadel, 148
Abu-Bakr, 41
Abu-Lughod, Janet, 160, 173, 176
Adams, Robert McC., 1, 2, 4, 8, 9, 11, 19, 159
Agha Khan, 68, 121
Agriculture, 2, 263
  arable land, 12–13
  aridity resistant crops, 18
  early, 129
  hydraulic, 131–34
  irrigation, 9, 19, 130
  produce, 8, 134
Alexandria, 164
Al Fustat (Egypt), 164
Ali, Caliph, 41–42, 60, 254
Ali'kosh (Iran), 17
Alliances, 244 (*see also* Social networks)
Al-Murra (Saudi Arabia), 125
Altaic (language), 89
Anatolia, 3
Animal Husbandry, 9, 19, 117, 181 (*see also* Domestication)
Antoun, Richard, 241, 246
Arabia Felix (Yemen), 6
Arabian desert, 7 (*see also* Rub'al-Khali)
Arabian shield, 4
Arabic (language), 24–25, 90, 109, 121
Arafat, Mount, 48
Aramaic, 98
Architecture, 18, 21, 163
Ardebil (Iran), 164
Armenian (language), 91
Armenians, 102–3
Aswad, Barbara, 199
Aswan Dam, 9, 131–32
Ataturk, 90
Ayrout, Father, 129
Azeez, A., 184

Babism (*see* Baha'i)
Baghdad, 165, 183–85
Baha'i, 104–5
Baluch, 111, 266
Baluchi (language), 92
Banuazizi, A., 234–35
Barth, Fredrik, 87, 121, 123, 253, 264
Basra (Iraq), 164
Basseri (Iran), 108, 116, 126–27, 264
Bates, D., 108, 116–17, 127
Battle of Badr, 40
Bazaar (*see* Markets)
Beaumont, Peter, 13
Beck, Lois, 124
Bedouin, 112, 234, 262, 263–64
Beirut, 185–86
Beni-Isad, 142
Berger, Morroe, 234
Berque, Jacques, 130, 157–58
Bezirgan, B. Q., 232
Bilingualism, 92
Bill, James A., 5
Bourdieu, Pierre, 217
Bradburd, Daniel, 124
Braidwood, Robert, 16
Buddhism, 31
Bujra, Abdala, 255–56
Byzantine, 24, 32, 33, 42, 60, 97, 108, 164

Cairo, 160–65, 173–78
Canaanite, 23
Catal Hüyük, 18

Chaldean, 98
Chatty, Dawn, 114
Childe, V. Gordon, 14, 15
Christianity, 23, 31, 97
Cilician Gates, 5
Circassian (language), 91, 106, 268
Circumcision, 50–51, 222, 245–46
Cities (*see* Urbanism)
Civilization (*see* State formation)
Class, social (*see* Social stratification)
Climate, 6–12
Cole, Donald, 114, 124
Confessional community, 84 (*see also* Sectarianism)
Constantinople, 164 (*see also* Istanbul)
Coon, Carlton, 3, 83, 258
Copt, 23, 94, 97, 174, 176
Costello, V. F., 133–34, 165, 187
Cotton, 133–34
Council of Chalcedon, 97
Cressey, G. B., 4
Cyrus the Great, 24

Damascus, 8, 164
Darbar, 123
Dariush Kabir Dam, 153
Dasht-i-Lut (Iran), 6
Dasht-i-Kavir (Iran), 6, 109
Dead Sea, 3, 4
Death duties, 31
Deciduous forests, 12
Dispute mediation (*see* Patronage)
Dome of the Rock, 50
Domestication of plants, animals, 15
Druze, 67, 68, 99

Economics:
  mixed economy, 17
  specialization, 270, 274
Edirne (Adrianople), 164
Education, 5–57, 202, 224, 228
Egypt, 4, 13
  arable land, 148–49
  distribution of land, 150
  early, 20
  fellahin, 129
  population, 13
  Saharan desert, 12
Elburz, 5, 111
El-Chebayish, 142
Empty Quarter (*see* Arabian desert)
English, Paul, 133, 165
Environment:
  physical, 1, 6, 131–34
  problems of, 1
Esposito, John, 233
Ethnicity, 84–88
  cultural diversity, 109
  ethnic mosaic, 109
Euphrates Dam (Syria), 70
Evans-Pritchard, E., 260

Family (*see* Kinship, Women)
Fatimid dynasty, 68
Fellahin (*see* Egypt)
Fernea, Elizabeth W., 232
Fernea, Robert, 141
Fisher, W. B., 1, 3, 13, 253
Flannery, Kent, 15, 16, 19

Geertz, Clifford, 273
Genealogies (*see* Kinship)
Geology, 3–6
Ghulat, 79
Gilsenen, Michael, 77
Gaziantep (Turkey), 117
Grazing cycles, 12
Gulick, John and Margaret, 196–97

Hadramout, 255–56
Halpern, M., 279
Hamidiya-shadhiliya (*see* Shadhiliya)
Hanafis, 53
Harik, Iliya, 237
Harlen, Jack, 16
Hashem, 38
Hebrew (language), 24, 95
Hijaz, 34
Hole, Frank, 16
Hopkins, Nicholas, 149
Hourani, Albert, 159
Households:
  compounds, 197
  extended form, 196
  kin, 218
  organization, 144–48
  resources, 9
  socialization, 220–24
  vengeance, 261
Hussain, 43, 63
Huxley, Fred, 245–46

Ibadis, 60
Ibrahim, Saad, 157
Indo-European (language), 89
Iranian Revolution (1978), 230, 252, 275
Irrigation (*see* Agriculture)
Islam:
  almsgiving, 46

Islam (*cont.*)
  conversion to, 59
  culture, 29, 81–82
  dietary rules, 50
  economic and political contexts, 32–34
  fasting, 47
  fatiha, 29
  Hijra, 39, 47
  jinn, 44
  law, 45–48, 51, 52–54, 65
  marriage, 35
  moral government, 250–52
  prayer, 29, 45, 46
  profession of faith, 45
  reformation, 49
  rise of, 32–34, 163
  rituals, 44, 48, 50–51
  sacrifice, 44, 48
  schisms, 60–62
  scope, 30–32, 250
  taboos, 50
  'ulama, 54–57
  Umma, 29, 42
Islamic calendar, 51
Islamic conquest, 25–27
Islamic revolution (*see* Iranian Revolution)
Isma'ilia, 67–68
Israel, 95, 271
Issawi, Charles, 275
Istanbul, 160, 161, 179–83

Jacobsen, Thorkild, 11
Jarmo (Iraq), 17
Jericho (Jordan), 17
Jerusalem, 165
Johnson, Gregory A., 20

Ka'ba, 34, 40 (*see also* Mecca)
Kaka'i order, 79
Kamal, Yashar, 139
Kashan, 134
Katkhoda, 122
Keban project (Turkey), 9
Kenyon, Kathleen, 17
Kerbala (Iraq), 165
Khamseh, 109, 121
Khamsin, 11
Kharijite, 61
Khomeini, Ayatollah, 43, 65, 273, 282
Khuri, F., 257–58
Khuzistan, 109
Kielstra, N., 246–47
Kinship, 40, 122, 145, 182, 189–95, 243–46, 258, 267
  classification, 191
  complementary opposition, 261
  descent, 206–8, 256
  duties, 198
  groupings, 198
  lineage, 126, 195, 260, 267
  matrilocal residence, 197
  patrikin, 262
  patrilateral, 197
  patrilineal descent, 194–95
  patronymic family, 195, 267
  rights, 195
  segmentary lineages, 261
  systems of address, 193
  terminology, 191
Kirman City (Iran), 165
Kolars, John, 118
Komachi (Iran), 124
Konya (Turkey), 165
Koran (Quran), 29, 38, 43, 52–54
Kramer, Noah, 22
Kurdish (language), 92, 109
Kur River Project (Iran), 10
Kuwait, 165

Labor (*see also* Migration):
  division of, 86, 274
  recruitment, 182
  sexual division, 224
  wage, 147
Ladino (dialect of Spanish), 91 (*see also* Sephardim)
Lakhmid, 25
Land:
  specialization, 155
  tenure, 131, 134–38
  use, 109, 124, 131–34
Languages, 89–92
Lapidus, Ira, 168
Law of family rights (1917), 229
Leadership:
  consensual, 42
  local, 249
  pastoral, 122
  power, 248–49
  religious (*'ulama*), 54–57
Lees, Susan H., 19
Lerner, Daniel, 234–35
Levant, 3, 8
Libya, 12
Lewis, Bernard, 24
Luri (language), 121
Luxor, 4

Magnarella, Paul, 192, 196, 203–4
Mahfouz, Najib, 177
Makhlouf, Carla, 233

Maliki, 53
Market economy, 113–14, 147
Markets, 166–68
Maronites, 99–102, 185
Marriage, 198–206, 247, 258, 260 (see *also* Women)
  arrangements, 201–4
  bride price, 181, 202
  bride wealth, 207
  endogamy, 199–200
  inheritance, 200
  levirate, 207
  monogamy, 200–201
  patterns, 198–99
  polygamy, 200–201
  post-marital residence, 205–6
Marvdasht Plain (Iran), 152–53
Marx, Emannuel, 114
Mecca (Saudi Arabia), 165
Medina (Saudi Arabia), 39, 40
Mehmet II, Sultan, 94
Mellaart, James, 18
Mernissi, Fatima, 215
Meshed (Iran), 165
Mesopotamia, 20, 23
Messianic belief, 64
Migration, 151, 171, 187
  exportation of labor, 277
  rural-urban, 226
Millenarianism, 65
Millet system, 94, 97, 100
Minerals, 3
Mizrachim, 97
Mocca (Yemen), 6
Mongol conquest, 25
Mountain systems, 3–6
Mu'awiya, 42
Mudif, 21
Muhammad, 34–39
Musaylima, 41

Nabatean, 24
Nadim, Nawal, 176
Nassar's Revolution, 238, 274, 280
Nelson, Cynthia, 233
Neolithic, 14, 16
Nestorians, 97, 98
Nile Valley, 3
Nizaris, 68
Nöldeke, A., 64
Nomadic herders (see Pastoralism)
Nusairis (Alawis), 67, 68

Oil (see Petroleum)
Omar, Caliph, 41
Osman, House of (see Ottoman)
Othman, 42
Ottoman, 26–27, 94, 99, 135, 142, 171

Palestinian refugees, 101–2
Plamyra, 24, 33
Pastoralism, 2, 13, 24, 33, 87, 107–28
  credit, 115
  ecological basis of, 108–10
  leadership, 112–14, 122
  livestock, 111
  military pressure, 111
  patterns of interaction, 118–19
  smuggling, 115
  state control over, 119–20
Patronage, 245–46
Peasant-nomad interaction, 110
Peasants, 129 (see *also* Agriculture)
Peristiany, J. P., 217
Persia, 108, 164
Persian (language), 90
Peters, Emrys, 194, 261
Petra (Jordan), 24, 33
Petroleum, 3, 4–5, 170, 186, 275
Phalangist (see Maronites)
Phoenicians, 23
Pitt-Rivers, Julian, 217
Politics, 116, 270
Population, 12–14, 19
Population density indices, 13
Prehistory, 14–20
Prohibitions, 23 (see *also* Islam)

Qaddiriya order, 73
Qash'qi, 116, 124
Qazvin (Iran), 164
Qom (Iran), 165
Quraish, 34, 36, 38, 40

Race, 88–89
Rainfall, 8–9
Ramadan, 47
Rassam, Amal, 219–20
Redman, Charles L., 16
Residence (see *also* Kinship):
  matrilocal, 197
  patrilineal, 208–9
  postmarital, 204–6
Rodinson, Maxime, 31, 35, 36
Romans, 33
Rub'al-Khali (Saudi Arabia), 109, 112,

Sa'adawi, Nawal, 232
Sadat, Anwar, 280

Safawi order (*see* Shabak)
Saharan desert, 12
Sakaltutan (Turkey), 144
Salim, Shaker, 142
Salzman, Philip, 115
Sarlyya order, 79
Sassanian, 24, 60, 133, 164
Saudi Arabia, 125
Saunders, Lucy, 130
Sayyid, 35, 63, 254–56
Sectarianism, 93–106, 184
Sedentarization, 114, 119–20, 121 (*see also* Pastoralism)
Seljuk Turks, 26
Semitic (language), 23, 89
Sephardim, 91, 95, 96
Settlement patterns, 12–14, 131–34
Sex roles (*see* Women)
Sexuality (*see* Women)
Shabak, 78–80, 87–88, 252
Shadduf, 11
Shadhiliya, 77–78
Shafi'i, 53
Shah, Idris, 71
Sharqi, 11
Shi'ite, 27, 54, 60, 62–69, 109, 185, 252, 283
  clergy, 65–67
  shi'at'Ali, (partisans of Ali), 41
  sects, 67–69
Sinai, 3, 8
Sirocco, 11
Sirs al-Layyan, 158
Slavery, 89
Social change, 78, 84
Social differentiation, 36
Social networks, 244
Social stratification, 46, 55, 122, 187, 255–57
Soil salinity, 8, 11, 132
Soysal, Mustafa, 134
Specialization:
  occupational, 86
  productive, 87, 107, 159
State control, 116, 143, 248, 264–67 (*see also* Tribalism and Centralization)
State formation, 20–23, 271
Stern, S. M., 162
Stirling, Paul, 140–41, 144
Sudan, 12
Sufi, 69–73, 75–80
Sumer, 23
Sunni Moslems, 50, 54–55, 109, 181
Suq (*see* Markets)
Surplus, 133

Tanta (Egypt), 165
Targum (dialect), 95
Tariqa, 70
Tatar (language), 91
Taurus, 3, 5, 11, 117, 128
Taxation, 42, 112 (*see also* Tribute)
Tenancy, 138–40
Trade, 3
Trade routes, 3, 112–13
Transhumance, 110–11, 117, 121, 125
Tribalism, 124, 250, 257–59
Tribe, 241–42
  confederation, 109
  organized village, 142–44
  patrilineal descent, 258
  structure, 259–63
Tribute, 112
Turkish (language), 90, 109, 121
Turkmen, 116, 202, 254, 259, 262, 263, 264

Umayyad Empire, 164
Urbanism, 19, 157–88, 264
  domination of region, 133
  location, 169
  squatters, 178–83
Urmia, Lake, 6

Van, Lake, 6
Village:
  headmen, 122, 140, 247, 249
  political organization, 140–45, 241

Wage labor, 128, 147, 151, 155, 225, 227 (*see also* Migration)
Wahabis, 75–77, 218, 267, 273
Warfare, 33
Water (*see also* Agriculture):
  availability, 111
  control and management, 9, 18
  resources, 131
  sources, 1
Watt, W. W. Montgomery, 35
Weiss, Harvey, 20
Wikan, Unni, 232
Wilkinson, J. C., 10
Wirth, L., 173
Women:
  bride theft, 222
  divorce, 223
  female circumcision, 222
  gamiya, 177
  haram, 214

hareem, 214
honor and shame, 217–20
labor, 227
menses, 222
modesty, 213–17
motherhood, 221–22
reproductive capacity, 215
role, 212–13
seclusion, 215, 224
segregation of, 223–24
sexuality, 214–17
status, 201
widowhood, 224
women's movement, 229
Wright, Henry T., 20

Yazid, 43
Yemen, 6, 255, 256 (*see also* Arabia Felix)
Yörük (Turkey), 108, 116–17, 127, 262

Zagros, 3, 5, 11, 128
Zaidis, 67, 69
Zoroastrianism, 25